Excursions
in and about
Newfoundland

During the Years 1839 and 1840

VOLUME 1

JOSEPH BEETE JUKES

CAMBRIDGE
UNIVERSITY PRESS

CAMBRIDGE UNIVERSITY PRESS

Cambridge, New York, Melbourne, Madrid, Cape Town,
Singapore, São Paolo, Delhi, Tokyo, Mexico City

Published in the United States of America by Cambridge University Press, New York

www.cambridge.org
Information on this title: www.cambridge.org/9781108030892

© in this compilation Cambridge University Press 2011

This edition first published 1842
This digitally printed version 2011

ISBN 978-1-108-03089-2 Paperback

CAMBRIDGE LIBRARY COLLECTION
Books of enduring scholarly value

Travel and Exploration

The history of travel writing dates back to the Bible, Caesar, the Vikings and the Crusaders, and its many themes include war, trade, science and recreation. Explorers from Columbus to Cook charted lands not previously visited by Western travellers, and were followed by merchants, missionaries, and colonists, who wrote accounts of their experiences. The development of steam power in the nineteenth century provided opportunities for increasing numbers of 'ordinary' people to travel further, more economically, and more safely, and resulted in great enthusiasm for travel writing among the reading public. Works included in this series range from first-hand descriptions of previously unrecorded places, to literary accounts of the strange habits of foreigners, to examples of the burgeoning numbers of guidebooks produced to satisfy the needs of a new kind of traveller - the tourist.

Excursions in and about Newfoundland

Joseph Beete Jukes (1811–1869) was a geologist who studied at Cambridge under the famous Adam Sedgwick (1785–1873) and eventually became a prominent member of the Geological Survey of Great Britain. In 1839, after many field expeditions in England, he was appointed to a survey of Newfoundland, a place about which he had until then been in 'utter ignorance'. The explorers failed to find the hoped-for mineral wealth they had been sent to prospect for, and returned to Britain. In 1841 Jukes joined the H.M.S. *Fly* as a naturalist for an upcoming expedition to chart the coasts of Australia and New Guinea. The *Fly* set sail for the Pacific in 1842, the year in which this two-volume account of Jukes' Newfoundland experiences was published. Volume 1 describes Jukes' arrival in Newfoundland, its rugged landscapes, and life in the fishing communities of this harsh North Atlantic outpost.

Cambridge University Press has long been a pioneer in the reissuing of out-of-print titles from its own backlist, producing digital reprints of books that are still sought after by scholars and students but could not be reprinted economically using traditional technology. The Cambridge Library Collection extends this activity to a wider range of books which are still of importance to researchers and professionals, either for the source material they contain, or as landmarks in the history of their academic discipline.

Drawing from the world-renowned collections in the Cambridge University Library, and guided by the advice of experts in each subject area, Cambridge University Press is using state-of-the-art scanning machines in its own Printing House to capture the content of each book selected for inclusion. The files are processed to give a consistently clear, crisp image, and the books finished to the high quality standard for which the Press is recognised around the world. The latest print-on-demand technology ensures that the books will remain available indefinitely, and that orders for single or multiple copies can quickly be supplied.

The Cambridge Library Collection will bring back to life books of enduring scholarly value (including out-of-copyright works originally issued by other publishers) across a wide range of disciplines in the humanities and social sciences and in science and technology.

EXCURSIONS

IN AND ABOUT

NEWFOUNDLAND,

DURING THE YEARS 1839 AND 1840.

By J. B. JUKES, M. A., F. G. S., F. C. P. S.;

OF ST. JOHN'S COLLEGE, CAMBRIDGE;

LATE GEOLOGICAL SURVEYOR OF NEWFOUNDLAND.

IN TWO VOLUMES.

VOLUME I.

LONDON:

JOHN MURRAY, ALBEMARLE STREET.

1842.

TO

CAPTAIN PRESCOTT, R.N., C.B.,

LATE GOVERNOR OF THE COLONY OF NEWFOUNDLAND AND ITS DEPENDENCIES,

&c.   &c.

DEAR SIR,

It would be an act of ingratitude on my part were I to publish any work relating to Newfoundland, and omit in some way to acknowledge the constant public support and the kind private hospitality with which you were so good as to honour me during my stay in that country.

Allow me, then, to dedicate this little book to you, as a slight mark of my respect and attachment; and in so doing to assure you that your kindness is gratefully remembered by,

Dear Sir,

Your very obliged and obedient friend and servant,

J. B. JUKES.

PREFACE.

I FEEL that the publication of a book made up of such slight materials as the following notes demands some apology. My reason, then, for sweeping out my note-books and laying their contents before the public is, the utter ignorance which prevails in England with regard to Newfoundland. Of this ignorance I must plead guilty to the possession of a full share before my visit to that country. Before I looked into books to learn something of the place I was going to, I did not know the name of the capital, or whether the island had any towns or permanent inhabitants or not. Since my return from the country I have found many persons—I may say, indeed, almost every one —in the same state of entire darkness on the subject. Very great ignorance of the condition and extent of our Colonial possessions, generally, prevails indeed among all ranks of society, except in those classes connected with them by trade; but Newfoundland, the oldest of all our Colonies, seems to have become utterly blotted out of our recollection, or is known only by its dogs. I thought, therefore, there might be a sufficient number of people willing to know something about it, and who would allow me to take them by the ear while I told them what I had seen of it. Scarcely had I begun, however, to enter on this task, when I was called to commence preparations for an excursion of some years, to another quarter of the globe. The marks of extreme haste, and many slips of the pen, will, no doubt, be abundantly evident to the reader, but let me assure him that such haste was the result of uncontrollable circumstances, and not a mark of disrespect towards him.

Should he desire to be more thoroughly and better informed on many points with respect to Newfoundland, I beg to refer him to a work shortly to be published by my friend Dr. Stabb of St. John's, which will, I believe, contain a full historical and statistical account of the island.

It now only remains for me to state that the Legislature of Newfoundland voted in their last session the sum of 100*l.* to defray the expense of printing my Report on the Physical Geography and Geology of the island, and lithographing and colouring its accompanying maps and sections. That vote, with the rest of the money bill, was subsequently lost in the Council. At my request, however, His Excellency the Governor, Sir John Harvey, has taken upon himself to pay this grant, and to him my acknowledgments are accordingly due, as enabling me to put everything in train before I leave England. I must here also publicly offer my thanks to the Honourable J. Crowdy, the Colonial Secretary, for the uniform courtesy and kindness I have received from him in this and other matters.

Wolverhampton,
February 8th, 1842.

CONTENTS OF VOLUME I.

CHAPTER V.

CHAPTER VI.

CHAPTER VII.

CHAPTER VIII.

EXCURSIONS

IN AND ABOUT

NEWFOUNDLAND.

CHAPTER I.

Departure from Liverpool—Falling in with Ice—Land in
St. John's—Visit to Portugal Cove, Cape St. Francis,
Shoal Bay, Bell Isle, and Harbour Grace.

April 11th, 1839.—Sailed from Liverpool in
the merchant brig "Diana," bound for St.
John's, Newfoundland.

May 4th.—The temperature of the air on
the open sea had been mild and pleasant, the
thermometer frequently rising to 60° in the
middle of the day; for the last day or two,
however, it had been gradually falling. To-
day, being in long. 48° W., lat. 47° N., there
was a thick fog, occasionally lifting and clearing off in open drifts in various directions, and
suddenly closing in again. The temperature

of the water fell rapidly as we proceeded on our course, and about two o'clock it was only 35°. The air, which at ten A.M. was 54, had likewise sunk to 38°. The swell of the sea gradually subsided, and the vessel was steadier than it had been since we sailed. A sharp look-out was accordingly kept, and as we were sitting after dinner in the cabin, about four P.M., the cry of "Ice a-head" brought us all on deck, and the vessel was put about immediately. I peered over the side into the driving fog, but could see nothing but small jagged pieces of white ice, hardly to be distinguished from the foam on the crest of a wave, whirling about in the eddy of the vessel. Beyond them was a kind of white obscure, that might probably be the sheen of the ice through the fog. Practised eyes, I suppose, saw more, and it was said to be the edge of a large field of ice, or rather to be the edge of "the ice," which was apparently understood to extend to the north for an indefinite distance. We soon left it far behind, and drove off to the southward. On returning to the cabin, the captain was congratulated on having made a good "ice-fall," and hopes expressed of as good a land-fall. Had we struck the ice in the night, we should probably have got entangled among it, and been greatly hindered and delayed on our voyage. During the two or three following days, large pieces of ice and small icebergs were frequently seen. Fogs also were frequent, occasionally clearing off entirely, and then recurring in a most singular manner. There seemed to me to be large well-defined banks of fog resting on the sea in various places, or moving slowly about, into and out of which we passed almost instantaneously, and which were not very easily observed by a person a little distance outside of them, with a bright sun overhead.

May 8th.—At sunset last night we saw the high land about Cape St. Francis, clearly defined on the horizon. At the sight of it, I seemed to myself, for the first time, really to have left the English shores. The vessel had unconsciously been associated in the mind with England, and appeared as it were a part of it: now, however, we looked on a distant land, and were really about to enter on a strange scene.

On coming on deck this morning we were

close in-shore, standing backwards and forwards across the mouth of the harbour, with a
light, but bitterly cold breeze blowing off the
land. The cold wind blowing over the ice the
other day was not to be compared to the
biting keenness of this, although the thermometer was a degree or two higher, being
about 41°. We were compelled, for the first
time since leaving Liverpool, to light a fire in
the cabin-stove, in order to breakfast in comfort.
The first view of the harbour of St. John's is
very striking. Lofty precipitous cliffs, of hard
dark red sandstone and conglomerate, range
along the coast, with deep water close at their
feet. Their beds plunge from a height of
from 400 to 700 feet, at an angle of 70°, right
into the sea, where they are ceaselessly dashed
against by the unbroken swell of the Atlantic
waves. This immense sea-wall is the side of
a narrow ridge of hills which strike along the
coast here, and through which there are occasional narrow vallies or ravines. These trans-
verse valleys cut down through the range to
various depths, and the bottom of one being
about fifty or sixty feet below the level of
the sea forms the entrance to the harbour

of St. John's, and is appropriately termed
the Narrows.   Inside, the harbour expands
and trends towards the S.W., and the land
on the other side of it has a much more gentle
slope, and a much less height than that immediately on the coast.   It is also of a better
quality, and more fertile.   The dark naked
rocks that frown along the coast near St.
John's, their stern outlines unbroken by any
other vegetation than a few stunted firs that
seem huddled together in the more sheltered
nooks and hollows, give a stranger but an
unfavourable idea of the country he has come
to visit, and seem to realize all the accounts he
may have heard or read of the coldness and
barrenness of the land.   As we sailed backwards and forwards across the mouth of the
Narrows, which in one place is only 220 yards
across, with rocky precipitous heights of 500
feet on each side, we caught a view of the
town, which, from its being built for the most
part of unpainted wood, had a sufficiently
sombre and dismal appearance.   The harbour,
however, was full of vessels, and on landing
there seemed to be much bustle and business
going on.   The melting of the previous

winter's snow had, however, furrowed the
streets in various places with gutters running
across them, while from their ill-kept state,
from their long, straggling, and irregular appearance, the narrow dirty alleys and lanes
leading out of them, the dingy aspect of the
unpainted houses, and the groups of idle
and half-drunken sailors and fishermen, the
absence of street lamps and drains, the
entire want of all police, and the air of dis-
order and confusion which reigned throughout, it was evident that the scene was a
foreign one. I found afterwards that it was
just the season when a number of vessels,
having shortly returned from the sealing expedition, their crews were all loitering about
with money in their pockets, and the merchants' wharfs and premises were crowded
with their men unloading the vessels, and
preparing the seals for the oil-vats. My first
impulse on landing was to ascend the ridge
on the south-east side of the harbour, which,
from the people all using compass bearings
instead of the true, is called the south side,
and the ridge the south side hill. From its top,
which is about 750 feet above the sea, there

was an extensive prospect over sea and land.
On returning to the town, myself and a fellow
voyager found it quite destitute of inns and
hotels, but were lucky enough to engage very
comfortable rooms in a private lodging-house.

May 9th to 20th.—Having called on the
governor and delivered my credentials, I
found myself most kindly and hospitably
received, not only by himself and family and
the authorities generally, but by a number
of private persons and families, to whom I
was gradually introduced. And here, once
for all, let me record my thanks to the many
kind friends I had the pleasure of making, for
their abundant hospitality, and assure them,
at the same time, that I shall studiously
avoid the ill-judged ostentation of gratitude,
of dragging their names before the public.
On the evening of May 12th, we were alarmed
by the cry of fire, and in a few hours a block
of houses in the principal street, in the
centre of the town, was burnt to the ground.
The behaviour of the military and the fire
companies was very good, and to their exertions and those of the more respectable inhabitants was owing the preservation of the

town. I was, however, much struck with the stupid indifference of a large part of the lower class of the population, as compared with the great, and sometimes self-baffling readiness and eagerness of the population of any large town in England, in similar circumstances. No inducement or excitement beyond that of present pay and reward seemed sufficient to rouse one of the hundreds of great idle fellows that stood around to stir hand or foot for the preservation of the houses and property about. I was afterwards told, indeed, that by far too many of the population looked upon a fire as a godsend, more especially if it reached or threatened a merchant's store, when a regular system of plunder was carried out unblushingly, and, as it were, by prescriptive right.

I think I afterwards perceived the causes of the low state of moral feeling* exhibited on this and one or two similar points, by people that in other respects had many excellent qualities.

---

* The former state of vassalage in which they were held by the merchants,—the adventurous nature of their pursuits leading them to look on danger or misfortune as necessary and inevitable,—as also the want of education, and community of feeling, and of a popular opinion, are among the causes alluded to.

During these first ten days I rode out
several times with the governor and the
surveyor-general, visiting Portugal Cove,
Logie Bay, and some other párts of the neighbourhood. Several roads out of St. John's
have been recently constructed, and are, for
about five miles, sufficiently good to ride or
drive on. The road to Portugal Cove is
finished from end to end, being about nine
miles and a half. The other roads beyond the
first five or six miles are either bare rugged
rock, wet moss and morass, or a bed of boulders, like the bed of a torrent. Sometimes a
combination of all three within the space of
twenty yards offers a choice of evils. Except
on a regularly constructed road, or over a
cleared and cultivated field, it would be quite
impracticable to take either horse, mule, or
ass a single yard. The country in the interior rises into many flat-topped or rounded
ridges, having an average height of 500 or
600 feet above the sea. One rounded hill,
five miles south-west of St. John's, called
Branscombe Hill, is nearly 900 feet high, and
from its summit there is an extensive pano-
ramic view of the land between the east coast

and Conception Bay. The land hereabouts was formerly nearly all covered with wood, consisting principally of various species of fir, spruce, and larch, with some birch. The wood is of a stunted growth, small pole-like trunks growing close together, with branches and twigs thickly interlacing from top to bottom : these, with the old trees falling across each other in all directions, form a matted thicket which is nearly impenetrable. In the immediate neighbourhood of St. John's, and occasionally along the line of the roads leading from it, these woods have been partially cleared, and those spots in which vegetable mould either was found, or could be formed, are turned into fields and gardens. The tops of the hills and ridges, more especially of the South Side Hill, are quite incapable of cultivation, being covered merely with dwarf berry-bearing bushes. Such spots are called " barrens." The hollows and slopes of the hills, and the bottoms of the valleys, contain generally either a pond or a marsh. A marsh is a tract of moss, which is sometimes several feet in depth, like a great sponge, and, of course,

always wet. The term pond is applied indis-
criminately to all pieces of freshwater, whatever may be their size; the grand pond on the
western side of the island being fifty miles
long. These ponds or lakes are inconceivably
numerous, the whole country being dotted over
with them. Hundreds of lakes, two or three
miles long and upwards might be found, while
those of a smaller size are incalculable. Immediately north of St. John's, and close to the
town, which is built on the low ridge that separates the harbour from it, is a pretty little lake
about a mile long, called Quiddy Viddy Pond.
About half way to Portugal Cove the road
winds for two miles along the banks of another,
called Twenty-mile Pond or Windsor Lake,
which with its surrounding woods is a characteristic and striking scene. On approaching
Portugal Cove the eye is struck by the serrated
and picturesque outline of the hills which run
along the coast from it towards Cape St. Francis, and presently delighted with the wild
beauty of the little valley or glen at the mouth
of which the cove is situated. The road winds
with several turns down the side of the valley,
into which some small brooks hurry their wa-

ters, flashing in the sunshine as they leap over
the rocks and down the ledges, through the dark
green of the woods. On turning the shoulder
of one of the hill-side slopes, the view opens
on Conception Bay, with the rocky points of
the cove immediately below, and the smooth
and fertile-looking Bell Isle in front at about
three miles distance, beyond which the high
lands of the other side of the bay were
seen still streaked and patched on their summits and sides with snow. On returning this
day (May 13th) from Portugal Cove, we were
overtaken by two or three smart snow-storms.

Logie Bay is a small indentation of the
coast, about four miles north of St. John's,
and is remarkable for the wildness of its rock
and cliff scenery : nothing like a beach is to
be found anywhere on this coast, the descent
to the sea being always difficult and generally
impracticable. In Logie Bay, the thick-bedded
dark sandstones and conglomerates stand bold
and bare in round-topped hills and precipices,
300 or 400 feet in height, with occasional fis-
sures traversing their jagged cliffs, and the
boiling waves of the Atlantic curling round
their feet in white eddies, or leaping against

their sides with huge spouts of foam and spray.
Just by Logie Bay is a mineral spring, containing among other things a large proportion
of iron in solution.   At the period of our visit
to this place, the scattered wooden huts were
empty and deserted, and the fish flakes and
stages out of repair, as the people only reside
during the summer season.

I had now been regularly installed as
Geological Surveyor, and instructed to examine into the structure of the country,
directing my attention in the first instance
to the neighbourhood of the principal towns
and inhabited places. A strong prejudice
existed that coal was to be found near St.
John's.   It behoved me, therefore, to find
out in what direction it could possibly lie.   I
shall not enter, however, on geological matters
in these notes, but refer the reader interested
in them to the Report in the Appendix.   My
first care was to engage a man who was acquainted with the country, as a guide and
servant.   Accordingly a rough-looking subject named Kelly, with a strong brogue, presented himself, whom, after some hesitation, I
engaged, or in his own language " shipped."

All domestic servants come to be " shipped."
Families are applied to to know whether they
want to " ship" a housemaid or a cook. Some
idea of the rate of wages may be formed by
the fact of my having to pay this fellow, a
common fisherman, 28*l.** currency, and his
board for the summer, that is, from the 1st of
May to the 31st of October. I found afterwards I paid him 6*l.* too much, but in strange
places one must always submit to a little im-
position by way of paying one's admission fees.

It will perhaps be the best way to afford the
reader some idea of the country, and of the
customs of the people, to give at some detail
a notice of the first two excursions I made from
St. John's, the one north to Cape St. Francis,
the other south to Petty Harbour and Shoal
Bay.

May 20th.—Dr. Stabb of St. John's having
agreed to accompany me to Pouche Cove,
where he had to visit a patient, we set off after
a good breakfast, at seven o'clock. Kelly car-

---

* 1*l.* sterling is equal to 1*l.* 3*s.* 4*d.* currency, the English
shilling passing for fourteen pence. The rule is, to turn
sterling into currency add a sixth ; to turn currency into
sterling subtract a seventh.

ried a knapsack and hammer-bag, while I was
encumbered with a prismatic compass, box
sextant, mountain barometer and thermometer, gun and shot-belt. There was a road as
far as Torbay, where I had been previously on
horseback : the last three miles of this road,
however, was merely a track over bare rock,
with all its original hollows and ruggedness.
On entering Torbay, we passed a small Roman Catholic chapel, which Kelly immediately
entered, and where he remained three or four
minutes. Torbay is a bold indentation of the
coast, offering little or no shelter except in the
coves at its head, and none there against an
easterly wind. The houses, as is always the
case in the small places, are scattered irregularly round the cove, with a straggling path
from one to the other. They are built of wood,
each by its own inhabitants, rarely painted, and
with small windows and doors, the chimney
and hearth being of brick. One or two have
two stories, and a second room or parlour. As
a shore frontage is of the first importance to
their fishing operations, every advantage is
taken to extend it. Zigzag stairs, supported
by short poles resting on points of rock and

wooden props, lead in some places down the face of the black slate cliffs, to a wild little landing-place on some sharp jagged ledges; while stages are often carried out on the rocks, affording places on which to spread the fish to dry. The greater part of the road we had come was bordered with thick scrubby wood: in one place, however, this had been burnt some years ago, and the white dry sticks and stakes still remained bare and brittle, crossing each other at all angles and positions, and as bad or worse to penetrate than the growing woods. As it was useless and almost impossible to work traverses or to find sections in such a country, I was driven to the sea-cliff. This, however, was inaccessible, except by a boat. I accordingly wished to engage a boat to take us to Pouche Cove, a distance of six or eight miles. I could not get one with fewer than four hands, and at the rate of 10s. a man, which, thinking extravagant, I refused. We then walked to Middle Cove, another part of the same bay, and there with some difficulty induced two men to pull us alongshore for an hour and back again in an old rickety leaky punt for a dollar and a half, or about 6s. and

3*d*. sterling. Returning to Torbay we got a dinner of tea, eggs, and bread and butter, the latter being common ship's biscuit and salt butter, for which we paid 4*s*. sterling. We then proceeded along the north shore of the bay by a little footpath on the edge of the cliff, and crossing over some high land came down on a small place called Flat Rock. Along the south side of this little cove, the thick red sandstone dips at a slight angle towards the sea, forming a long, smooth, sloping pavement, whence the name of the place. At this point we left the shore, and following a narrow path into the woods, came to a brook crossed by a wooden bridge. This bridge consisted of two or three beams laid across the stream, with a number of poles nailed down across them side by side. Just beyond this, a part of the woods had been recently burnt: the black charred trunks and branches of the trees still standing among the heaps of ashes and charcoal had a striking and singular appearance when viewed for the first time. A path several feet wide had been cut through the woods for three or four miles beyond this, but no other pains had been taken with the road than in a few of the

more boggy places to lay down some sticks
and branches of trees, by means of which a
dexterous jumper leaping from one to the other
might avoid sinking much more than ankle
deep. The slippery sticks, however, made a
fall so often inevitable, that, to avoid risk to
the barometer, I was content to wade how I best
could. We passed one or two little ponds of a
few hundred yards in diameter, embosomed in
wood. At the last of these the scenery was
very characteristic of a wild country. The tranquil pond seen at a little distance through the
trees, with the thick rank forest circling it on
every side so as to exclude every breath of
air, old trees covered with tangled bunches of
moss and white lichens, fallen through sheer
decay and leaning against their neighbours in
all positions over the water of the pond, presented a scene which had evidently been un-
altered for centuries, and which no act or care
of man had ever modified in the least; while
from the full margin of the pond a small brook,
some five yards wide, leaped with sudden im-
petuosity, and foaming over a rocky bed at our
feet, hurried into the woods, frequently concealed by masses of fallen timber from bank to

bank, utterly lost to sight in about thirty or
forty yards, and only to be seen again when,
in sailing alongshore, it was found to fling its
waters over the broken crags into the sea.

At this point the broad path ended, and a
little narrow winding crevice, as it were, cut
through the woods, barely sufficient in some
places to allow a broad-shouldered man to
squeeze between the crowded stems of the pole-
like trees conducted us to Pouche Cove. It was
just dusk when we arrived, and we put up at
the first house or cabin we came to. The people
received us most hospitably, and gave us tea
and bread and butter, but, owing to the recent
death of a daughter, could not accommodate
us for the night. We then were taken to the
house of the schoolmaster, where we were
kindly received; but his house being equally
unfurnished with room for more than his own
family, he took us to a Mrs. Sullivan's, where,
it appeared, strangers usually put up. We
found here several people assembled round the
wood-fire, and shortly joined the circle. After
some interchange of talk, in which Kelly bore
his part by retailing all the news of St. John's,
Dr. S. and I were shown through a door into

a small narrow room, in which there were two
beds. I, in my ignorance, concluded this was
a bed a piece; but Dr. Stabb, more accustomed
to the country, immediately asked who slept
in the other bed, " Myself and the girl, sir,"
said the venerable Mrs. Sullivan, to my great
astonishment. Accordingly we tumbled into
one bed, and after the fatigues of the day were
soon fast asleep, and in the morning found the
other bed had certainly been slept in, and so
concluded the old lady and the girl had effected
their entrance and exit quietly in the night
without disturbing our slumbers.*

May 21st.—The bake-pot was put in requi-
sition this morning for our service, and a lot

---

* This was the only instance I met with myself in the
country in which what we should consider the rules of de-
corum were violated. I heard frequent anecdotes, however,
of the patriarchal simplicity of the habits of the people a
few years ago. It was a common thing, for instance, for
the man and his wife to admit the solitary wayfarer to a
share of their bed : but it must be recollected that in the
severe winter of the country this was probably necessary for
the preservation of his life during the night; and where the
whole house consisted but of one room, as was often the
case, separation was of course impossible. Among a people
of simple habits, however, sleeping in one common room is
not looked upon as more indecorous than sitting, eating,
or drinking in it.

of little cakes made by our hostess for breakfast, at which she gave us also some fresh
herrings. While this was cooking we amused
ourselves by watching the spouting of two
whales in the offing. I then wished to engage a boat to pull round Cape St. Francis and
down to Portugal Cove, but after considerable discussion I could not get one at any but
what I thought an extravagant price, 3*l*. being the lowest sum asked. I accordingly took
leave of Dr. Stabb, who wished to return to
St. John's, and intended to make some excursions on foot, and if possible to force my way
to Portugal Cove by that means. As I first
wished to see the rocks on the south side of
the cove, I struck through the woods in that
direction. The distance was about a mile, but
it cost an hour's scramble, in the course of
which both Kelly and I were wet through,
from rain which fell early in the morning and
still rested on the leaves, to get near where I
wished. I then found a deep narrow " gulch "
or crevice in the rocks still separated us from
the point of the cove, and that it would cost
another hour at least to reach it. A boat was
passing beneath, but a look at the smooth

shelving ledges and black precipitous slate cliffs at once forbade all attempt at a descent. We accordingly returned to Pouche Cove, and passing through the scattered lot of straggling houses, proceeded towards Cape St. Francis. A narrow winding path of three miles through the dense woods conducted us to it. Just as we emerged from the woods on a bare eminence above the little cove behind the cape, called Biskin or Biscayan Cove, a schooner rounded the cape close in-shore, apparently nearly touching the rocks, and then, hauling her wind, kept close along the coast for St. John's. The lively craft, contrasted with the deep silent woods and the wild rocks of the cape, formed a beautiful picture. A few fishermen's huts were scattered about even in this desolate-looking little place. I walked round Cape St. Francis, admiring the boldness of its rock scenery and the magnificent sea-view of the mouth of Conception Bay, but soon found all further progress hopeless. It was not until I actually looked upon this coast with my own eyes that I could fully appreciate or even understand the difficulty, if not the absolute impossibility, of traversing it. It is truly an iron-

bound coast. It is not only lofty, and frequently
so precipitous as to be nearly perpendicular or
even sometimes overhanging, but so broken
by frequent indentations and precipitous crevices and ravines, as to afford no regular edge
or summit along which to walk. Meanwhile to
the very verge of the rocks, and wherever there
was a possibility of a sufficient thickness of
moss supporting itself, the compact array of
stunted fir-trees crowded themselves, as if re-
solved to bar all thoughts of a passage. I
now began to understand the quiet smile of
ridicule with which the people met my propo-
sition to walk across from Pouche Cove or
Cape St. Francis to Portugal Cove, and the
decided refusal they gave to be partakers in
such an extraordinary undertaking as it evidently appeared to them. I could not before
conceive what possible obstacle there could be
to render a journey of only six or eight miles
such a mighty matter; and was astounded to
hear it related as rather a doubtful fact, that
one or two men had sometimes done it in the
winter when the snow had got hard-frozen and
rendered the ponds and marshes capable of
bearing their weight, and thus enabled them

to avoid great part of the woods. When I recollected that from the top of Branscombe Hill,
five miles to the south-west of St. John's, I
could see no termination to these woods, my
idea of geologizing the country by the usual
method of a series of traverses, certainly received a very rude shock, and I began to perceive that some other plan more adapted to the
nature of the country must be adopted. We returned, however, to Biskin Cove, and got some
dinner of tea, eggs and herrings ; and I then
agreed with three men for three dollars, or 15s.
currency, equal to 12s. 6d. sterling, to row me
back to Pouche Cove, a distance by water of
about three or four miles. The rock scenery
of these black cliffs, from two to three hundred
feet high, with the white surf foaming in their
crevices, was certainly very fine, and as we
rose and fell with the swell beneath them, assumed a stern grandeur of appearance still
higher than their real character. On landing
at Pouche Cove I observed what I had not
before remarked, namely, slides or ways made
of trees and rough planks in one or two crevices of the cliffs, and found they were for
dragging up the boats and skiffs whenever an

easterly gale set in, as they were then no
longer safe at anchor: some of these slides
were forty or fifty feet high, and very steep.
As soon as we left the boat the men went up
into the village, and Kelly said they would
spend all the money in rum before they went
home. We set out for Torbay by the same
road we traversed yesterday, Kelly's tongue
running on in spite of the difficulties of the
path : among other things he said, speaking
of the recent establishments of schools in the
outports, " What a fine thing it is, sir, to have
them schoolmasthers erected in every part of
the country ! " Nothing remarkable happened
in our return, except that I felt absolutely faint
with hunger before we reached Flat Rock, in
spite of the dinner I had eaten at the Cape.
Kelly's solution of this phenomenon was, that
there was a kind of grass they called hungry
grass, and whoever passed over it immediately
became so faint for want of food, that unless
they could shortly obtain it they would drop
and perish by the way. On this account it was
an established rule never to go even a few
miles into the country without a cake of bread
in the pocket, at least. I fancied the cause of

my hunger was, that my English stomach, having been accustomed to beef and ale, disdained to be satisfied with tea and fish in whatever quantity supplied, especially with an appetite sharpened by keen air and exercise. However that may be, I attacked the first house I came to, where we luckily got a quart of milk, which, with some bread and butter, carried us to Torbay. Here we put up at a larger house than common, where I got a little salt mutton, and a decent bed and bedroom with a parlour attached, signs, I suppose, of our approach to the capital. I had nearly given up smoking, but this evening felt a sudden desire for tobacco. I accordingly procured an old pipe and some strong negro-head tobacco, and certainly found by its use a great relief from the fatigue and over-excitement of travelling, and took care afterwards never to be without it, especially where there was a probability of short commons. I paid here a dollar and a half for our supper and bed, and returned the next morning to St. John's to breakfast.

May 23rd.—About three in the afternoon I set out for Petty Harbour, intending to visit a place about five miles beyond it where

there was reported to be a copper-mine that
had been worked in the last century. For the
first three or four miles there was a good road,
but we then turned off by a narrow path over
the ridge of the South Side Hill. The summit
of this is a broad bare moorland, consisting
partly of morass or small skirts of wood, partly
of naked sheets of level rock or of round hummocks and knobs, and small abrupt ridges of
rock rising up here and there along " the
strike " of the beds. There are several ponds
or small lakes scattered about it : one or two of
these are connected by narrow passages, and
form most picturesque sheets of water ; and
when the eye stretches across them over the
well-defined boundary of the hill just beyond,
and sees far away in the distance the blue
horizon of the sea, few scenes that I have be-
held are more wild and striking. After traversing this moorland, we came suddenly on
the verge of its seaward slope ; and there in a
narrow ravine between dark precipices lay the
cluster of white houses called Petty Harbour.
The houses surround a small creek, which re-
ceives a howling torrent that hurries over the
rocks of a desolate valley just behind, and they

c 2

seem so secluded and shut out from the world,
and the people too seemed so well off and contented, that I was much interested with the
place altogether.   There was a small inn also,
where I got very decent quarters.

May 24th.—At six A.M. I started in a four-
oared punt * for Shoal Bay, about five miles
down the coast.   The morning was cool, but
beautiful, with a light air off-shore; and the
bare lofty precipices of hard red sandstone,
which here form the coast, produced magnificent cliff scenery.   There was of course
a swell, as we were on the margin of the wide
Atlantic, and the manner in which we effected
a landing was by rowing on the crest of a
wave into a crevice just wide enough to admit the boat, and which shortly turned round
at right angles behind a broad mass of rock,
forming thus a snug shelter for our little craft.
After some search we found the place where
the mine had formerly been, and where some
iron staples and bolts still remained in the
rocks.   There was, however, no appearance of
a shaft, nor could I find anything which could
lead me to guess as to the size or importance

* Punt is the term usually given to a small row-boat.

of the vein. Some pieces of vein stuff, apparently old refuse, lay about, containing small
patches and strings of ore, which proved to be
grey sulphuret of copper. After a delay of an
hour or two the wind gradually rose, and,
seeming inclined to shift towards the east,
warned us to be off, which warning it was
lucky we obeyed, as the sea was rising, and
by the time we returned to Petty Harbour it
was blowing pretty strong. On our return we
saw many fishing-boats, and passed one that
was anchored under a headland, jigging cod-
fish. A jigger is a plummet of lead, with two
or three hooks stuck at the bottom, projecting
on every side, and quite bare. This is let
down by the line to the proper depth, and
then a man, taking a hitch of the line in
his hand, jerks it smartly in, the full length of
his arm, then lets it down slowly and jerks it
in again. The fish are attracted by seeing
something moving in the water, and every
now and then one is caught by one of the
hooks. As soon as the man feels he has
struck one, he hauls in upon the line, taking
care to keep it tight till he heaves the fish into
the boat. In this way several were caught

while we were in sight by the two men in the
fishing-boat, and they tossed us three or four
fine fish as we went by, at the request of the
owner of our punt. Returning to Petty Harbour, I walked back to St. John's by the way
I had come yesterday.

May 29th.—I was at Portugal Cove, and at
eight A.M. started in a punt with four men to
visit Bell Isle. It was rather a cloudy morning, but clear when we set off. After a row of
about three miles we landed on the beach,
which juts out under the cliff for about a
quarter of a mile in length and a hundred
yards in breadth. Found the island to consist
of black shale, with interstratified beds of grey
gritstone, very different from the hard slates
of the mainland opposite. Except at the
beach, the whole island is bounded by perpendicular cliffs, at one point nearly 300 feet
high. As I stood on the verge of this highest
point, small streaks of thick fog began to steal
along the surface of the water beneath, and
eventually covered it from sight, except here
and there, where a transient opening showed
a patch of grey water dotted with flakes of
foam. Returning to the boat, we went south-

about and made the circuit of the island, which is about five miles long and nearly three broad. Near the south-east extremity is a very pretty little place called Lants Cove, in a small indentation of the cliff, this and the beach being nearly, if not quite, the only points at which it is possible to land. Both these points contain several houses and inhabitants. On the south-west side of the island we passed between it and the Bell, as it is called, a perpendicular rock about an acre in extent, whose flat top is level with the adjacent land, being more than 100 feet above the water : a narrow strait of about twenty yards across admits of the passage of a boat in calm weather. Having passed through this, we had to row up the west side of the island against a headwind and swell and thick driving fog. There was still, however, a fine scene, to one unaccustomed to the sight, in the dim looming through the fog of each precipitous headland as they successively appeared, frowning nakedly above us as we passed beneath their feet, and then seeming to wrap themselves in cloud again as we laboured slowly onwards. The chill, however, of the fog, and the rolling of our

little boat in the swell, gradually produced
a state of body very unfavourable to the appre-
ciation of the picturesque, that of incipient sea-
sickness, and I was heartily glad when we
again landed at Portugal Cove. Here the sun
was shining brilliantly, and I thought the day
had changed remarkably, but on looking back
was surprised to find Bell Isle invisible, and
that the whole of the bay was enveloped in
dense white fog rolling in before a north-east
wind, while the hills and the land around us
were basking in the sunshine without a cloud
or a mist.

I heard this evening that they had found
coal at Harbour Grace, and that the little
packet-boat which plies across the bay had
brought specimens of it, which had been taken
to St. John's. Accordingly, the next day I
hurried across in the packet to Harbour Grace,
expecting to find twenty or thirty men digging
away at the outcrop of some bed of coal. On
inquiry, however, no one seemed to know
anything about it, till I was taken to a black-
smith's shop, who said he had got coal from
the banks of a lake called Lady Pond, which
he had used in his forge and found it answer

admirably. A lad accordingly undertook to
show me the place, and, the affair getting wind,
I proceeded, under a strong escort of twenty or
thirty people, in search of the coal. At the
back of the town I came upon a ridge of hard
clay slate, thick bedded, with imperfect cleav-
age, which rather shook my belief in the
neighbourhood of a coal-field; but as in a new
country there might be new facts, I did not
despair, and thought that at least a bed or a
nest of anthracite must exist. After pushing
through the woods for some distance, we came
to the margin of a lake a mile long, with a
perfectly flat shore, and without an inch of
rock exposed, and my conductors began
dabbling in the water till they picked up some
little brown pieces of stuff like a light cinder,
about half the size of an alderman's thumb,
and which they strenuously persisted was coal.
One young fellow, however, said that sleighing
parties often made fires on the ice near this
spot in the winter, which confirmed me in the
belief of their being cinders. However, I took
some pieces away with me, and afterwards, on
drying and testing them, found them to be

c 3

pieces of a light spongy variety of bog-iron
ore. So ended all my hopes of a Harbour
Grace coal-field, but still so strong and so
general was the belief of coal existing in the
neighbourhood, that I thought there must be
some ground at least for it, and therefore
determined to explore with sufficient diligence
to decide the question. But how was I to explore? To go hunting through the woods
and traversing the marshes was an idle waste
of time, and I found that in the approaching
fishing season it would be impossible to hire
boats: I therefore came to the determination,
at my return to St. John's, to report to the
governor, with a request to be furnished with
a small coasting-vessel, in order to examine
the cliffs in detail. To return, however, to
Harbour Grace. It is a pretty-looking little
town, consisting of one long straggling street
along the north side of the inlet or harbour,
the houses being mostly painted white, and
standing on a narrow flat with a rocky ridge
just behind them. Its population is about
3000. It has altogether a more English and
neat appearance than most places in New-

foundland. It contains, moreover, a very
decent inn, which at this time even St.
John's was destitute of.

May 31st.—Walked across to Spaniard's
Bay. The first three miles there was a good
road, but for the remainder of the distance, or
about three more, it was merely marked out.
From the summit of the ridge, between Harbour Grace and Spaniard's Bay, the view
was highly picturesque. The foreground
consisted of rocky eminences covered with
wood, with lakes of water in their hollows,
beyond which the eye looked over a succession
of land and water formed by the inlets of
Spaniard's Bay, Bay Roberts, and Port de
Grave, and the narrow craggy necks of land
between them, with the Cats Cove Hills, and
the other highlands about the head of Conception Bay, in the distance. Spaniard's Bay
contains but a few houses, but, as a gentleman
resided there who was a fellow-voyager with
me on my passage out, I stayed and spent the
day with him. I found everywhere the same
hard useless sort of slate rock as about Harbour Grace.

June 1st.—Returned early in the morning

to Harbour Grace, and got on board the little sailing-packet to return to Portugal Cove. As there was a fine north-west wind blowing, I thought it needless to lay in stock for a voyage of only sixteen miles, but I soon found my mistake. We had hardly got well out into the bay before we were becalmed, and gradually drifted alongshore by the current; and, after a tedious passage of nineteen hours, only landed in Portugal Cove at half-past four the next morning. Had it not been for the kindness of two gentlemen going to St. John's, who were better provided than myself, I should have had but a very indifferent time of it.

CHAPTER II.

Engage the Beaufort — Visit to Harbour Grace, Cupid's
Cove, Brigus, the Cat's Cove Hills, Holyrood Butterpots,
and Kelly's Island — Visit to Carbonear, Bacalao Island—
Ice Islands off Trinity Harbour—Visit to Trinity, Random Sound, Come-by-Chance, Chapel Arm, New Harbour, Heart's Content—Return to St. John's.

THE Legislature of Newfoundland had originally voted 350*l.* towards the geological survey of the island. In consequence, however, of
my report to the governor, they increased the
grant this year to 600*l.*, in order that I might
procure a coasting-vessel with which to survey
the shore in detail from point to point. After
some trouble I engaged a small ketch, called
the Beaufort, about thirty-seven tons burthen,
belonging to a man named Gaden, who was to
go himself as master. He was besides to get
a good four-oared punt, four men, and a boy,
and to victual himself and crew. For this I
engaged to pay 55*l.* per month. I had then to
furnish the cabin, which, for the size of the

vessel, was large and convenient, (as I could
stand upright . under the skylight, and had
room for a chair, a stove, and a small fixed
table,) lay in stores for my own provision, as
well as a stock of everything I could possibly
stand in need of. All these preparations being
completed, she was ready for sea on June 15th,
when I slept on board.

June 16th.—At five in the morning I was
awoke by a noise on deck, which I found to
proceed from one of the sailors being brought
on board drunk, having after some search
been caught by Gaden in the streets. It was
eight o'clock before we were fairly under
weigh, and stood alongshore to the northward
before a light air, so light indeed that it was
evening before we floated round Cape St.
Francis, and entered Conception Bay. At
ten P.M. there was a red gleam of sunset
fading in the west, with a fine arch of aurora
and flickering streamers over it, while above
was a young moon and all the host of stars.
The shores of the bay were visible on either
hand, the bold coast between Cape St. Francis
and Portugal Cove standing in dark relief
against the clear star-studded sky.

June 17th.—It was quite calm during the night, but at six A.M. a breeze sprung up from the south-west, against which we beat up for Portugal Cove. As we neared Bell Isle we saw a fleet of fishing-vessels rounding it and anchoring under its lee, and presently a smart squall of wind and rain struck us, against which we made but little headway. After trying for some time in vain to get near enough to Bell Isle to anchor, we bore up for Harbour Grace. The packet-boat put out of Harbour Grace as we went in, but after a little while she too ran back to the shelter of the harbour.

In the afternoon I visited a land-slip, in company with Mr. St. John. In June, 1838, near a place called the Grove, on the north side of the harbour, after a great flood of rain, a large piece of the cliff, consisting chiefly of rubbish and boulders, slid forward over the smooth slate rock, which in that place dips towards the water. The subsidence was about 100 yards long, 40 yards across, and the cliff about 140 feet high. It consisted now of three ridges or shelves of land on which the trees stood undisturbed, with broken ground between them. These shelf-strips rose one above another, the uppermost having subsided about fifteen feet, with a chasm thirty feet wide between it and the main land.

The next day I saw some pretty good slate-rock, on the banks of the brook at the head of the harbour, but found the woods very difficult to penetrate. On returning by the shore of the Lady Pond, I saw something struggling in a shallow place, and fired at it. A large eel showed himself, and I gave him the other barrel and stunned him. One of the men went and got hold of him, but loosed him again for fear he should *be bitten;* he then stooped down and seized him with his teeth and both hands, and threw him into the air, but instead of sending him landwards heaved him farther into the water, so we lost him. He was very large, and I was annoyed at the man's stupidity.

June 19th.—Weighed anchor at seven A.M., and, in going out, visited the small island (one of a cluster of rocks) at the mouth of the harbour, on which stands a lighthouse, perched

on the brink of a precipice, consisting of a
square wooden house with a square dovecote-
looking top for a lantern.   It is, however, an
effective and useful light.   The island is only
accessible in tolerably smooth weather, by
ladders up the face of its landward cliff.   It
consists of one large block of as fine roofing-
slate as ever I saw in my life.   Then went in
the boat to Bryant's Cove, a picturesque little
spot, being the mouth of a small valley that
here comes out on the sea, and then proceeded to Spaniard's Bay, whence I sent the
boat to tow the vessel in, it having fallen quite
calm.

June 20th.—This day visited Bay Roberts
and Port de Grave, finding nothing but slate
rock, similar to that which I had seen before.
Both these places are inhabited.   Indeed the
western (or as it is called in Newfoundland the
northern) shore of Conception Bay is the most
populous part of the island, having upwards of
20,000 people scattered along it, with ministers
of the churches of England and Rome and of
the Wesleyans, several medical men, and most
of the comforts and conveniences of civilized
life.

June 21st.—As I wanted to get a little into
the interior, to see if there were any boundary
to this abominable slate rock in that direction,
we took the boat this morning to what is
called the northern gut of Port de Grave.
Here a brook empties itself into the sea, having
run for about three miles through a narrow
pond, or "cosh," as my men called it. We
hauled the boat up the shallow bed of the
brook for a few yards into deeper water, and
then rowed up the lake. I then struck off,
with two men and the theodolite and barometer, to the top of a hill on the north side,
called Sunday's Hill, whose height I found to
be about 500 feet. From its top I could see
nothing in the interior but woods and marshes,
dotted with ponds, with several rugged eminences rising in different directions. I proceeded two or three miles farther, over some
marshes, till I found the labour as useless as it
was great, since no rock showed itself. We
then had a hard tug through some thick
woods down to the bed of the brook, where
I found still the same slate rock, and after
much toil and difficulty we got back to the
head of the "cosh." Here we met Gaden,

who had caught three or four dozen of trout
up the brook, but on coming to the boat found
some urchins from the neighbouring houses
had stolen all our bread and butter. So we
sailed down the cosh to a house, where we
procured some potatoes, and with 'these and
the trout we had caught made a comfortable
dinner. As the evening fell I took the rod
and line, and just where the brook ran into the
sea I caught several dozen very nice trout and
salmon-peel. They seemed both to prefer the
brackish water at the mouth of the brook, and
the largest were caught farthest into the salt
water. We then returned to our vessel, which
was anchored about a mile down the bay, in a
place called Cupid's Cove, but which, contrary
to what might have been expected, was a wild
desolate little spot with no attractions whatever. I was much struck with the beauty of
the little valley we had visited, its sheltered
situation between two bold rocky ridges, and
the apparent superiority of the soil on the
banks of the pond and brook, as evinced by
the larger size of the trees, and the little
patches of natural grass or turf which were
to be seen. Of the latter the generality of the

country is quite destitute. The seafaring pro-
pensities and employments of the people, however, had hindered all but just the seaward
end of this pretty little place from being taken
possession of.*

June 22nd, 23rd, 24th.—Proceeded to Brigus, and examined its neighbourhood. This
is a rather considerable settlement, with a
population of about 2000. It is a wild rocky
little place; but about three miles inland from
it is a fertile valley, through which runs a
brook forming occasional ponds, and emptying
itself into the sea on the southern side of Port
de Grave. It is accordingly called the Southern Gut. In this valley lies a farm called
"the Golds,"† which was farmed by Mr.
Cousins. The road to it lay through a pretty
wood, in which the timber was greatly superior
to that commonly to be found. There was

---

* Since the new road has been completed, it has several
times struck me that it would be worth while to clear away
the bar at the mouth of the brook, convert the "cosh" into
a small harbour, and cultivate its banks on a larger scale. I
know few places that would exceed it in beauty as a summer residence, and for feeding cattle it would certainly be
excellently adapted.

† So called from a yellow flower which grows abundantly on the banks of some of the brooks.

some good grass-land in the flat, but much of
it had been destroyed by spring floods, as the
best portion of it was scarcely above the level
of the brook. About two feet below the surface
was a stratum, three or four inches thick, of
concretionary bog-iron ore, that had as usual
been mistaken for coal. The water, becoming
impregnated with iron, probably from the
decomposition of sulphuret of iron in the slate
rocks, deposits it when it becomes stagnant in
the flats, and it forms concretionary nodules in
the clay or mud. Much of this flat land was
covered with raspberry-bushes; and Mr.
Cousins informed me that, after a fire in the
woods, the first thing that covers the ground
is a luxuriant growth of raspberry-bushes,
which are gradually succeeded by a thick
wood of birch, although previous to the fire
nothing but fir and spruce may have been seen
for miles. I believe that at the time of my
visit this farm was not a profitable speculation,
although the only one in the neighbourhood.

June 25th.—I set out this morning in the
boat, accompanied by Mr. Green of Brigus, to
visit the Cat's.Cove Hills at the head of Collier's Bay. After-exploring the coast, we arrived

at the head of the bay at 12 o'clock, and
struck off into the woods by a narrow path.
We took three men to carry the theodolite and
hammer-bag, but I did not intrust the barometer to any one, and accordingly carried it
myself. In about a quarter of a mile we came
upon a newly-cut road that was intended to
run from Brigus round the head of the bay
towards St. John's. We travelled along this
for about half a mile, but found it a mere morass, and were obliged to jump from one root
of a tree to another the greatest part of the
way. We then struck off across a marsh
for the hills, and in a short time entered a
wood. Here we found the bed of a little torrent, which enabled us to proceed some distance
with tolerable ease, but when that ended we
were obliged to force our way through the
dense thicket, sometimes climbing over, some-
time crawling beneath, masses of fallen rotten
wood, stumbling over slippery moss-covered
boulders, slipping on the wet roots of trees,
sliding down steep banks of rock, or tearing
the clothes off our backs by dragging ourselves through the matted twigs and branches
of the trees. All this while the thermometer

stood at 75°, and not a breath of air could find its way to us. The smell of the woods and the turpentine exuding from the trees was as close and stifling as an oven, and the mosquitoes began to attack us by hundreds. They did not, indeed, annoy me so much, but Mr. Green's face and neck were soon running down with blood. When, after toiling in this way to the top of one ridge, we still found a ravine separating us from the hill, we were half inclined to give it up. However, taking " heart of grace," we plunged again into the woods, and after another stiff climb we at length found ourselves on the summit. It cost us altogether more than four hours' hard labour, although the distance was not much more than three miles from the sea-side. The hill consists of three precipitous peaks, one of which seemed nearly inaccessible, but, notwithstanding its rugged and mountainous appearance, the barometrical observations did not give it a greater height than 860 feet. The view from its summit was very beautiful : the whole of Conception Bay, with its islands and varied shores, was spread out before us on one side, and all the centre of the province of Averlon on the other. A tract of comparatively flat country stretched away to the south towards St. Mary's Bay, varied, however, by undulations and occasional eminences, and covered with woods and lakes, one of the latter of which seemed of considerable size. To the east of this tract ran a line of round hummocky hills, consisting apparently of bare rock : to the west was a ridge of hills seemingly of greater height, but of a rounder and smoother character. After spending an hour on the summit, and taking a round of angles on the principal objects, we descended. In going down we took care to choose a better route than the one we came up by, and, after traversing a small marsh or two, came upon a wood-path, by following which we got out to the sea-side in half the time it took us to go in. These wood-paths are frequent in the woods near the sea-shore : they are formed in the winter by the people coming to cut fire-wood, when they penetrate in various directions in search of the best timber, and haul it out over the snow and ice, having previously cut a crevice through the thicket wide enough to admit its passage. In the summer, when the snow is off the

ground, these paths are sufficiently rugged
and uneven, as the stumps and roots of the
fallen trees still remain; they are, however, a
little better than the uncut woods, and serve at
least to keep a person in the right direction, by
which much time is saved. As they are intended to be traversed only in the winter, however, they invariably cross all the marshes and
ponds that lie in the way, and the traveller is
constantly stopped by coming suddenly on a
large pond, right on the other side of which
he sees the continuation of the path, with the
pleasing consciousness that the space which
would require ten minutes' easy and pleasant
walking, were the pond frozen over so as to
enable him to cross, will cost him now an
hour's hard and provoking labour before he
can get round it. It is, moreover, round the
edges of the ponds that the trees are most
abundant, and the underwood thickest and
most impervious. On getting out to the sea-
shore we got some bread and butter and molasses-tea,* at a house or hut hard by, where a

---

* Common tea, *boiled* in a tea-kettle, with a spoonful of
molasses for sweetening, and often drunk, in default of a
better cup, out of the tea-kettle lid.

family had resided during the winter, and returned to Brigus by ten o'clock.

June 26th and 27th.—Sailed along and examined all the inlets and holes and corners round
the head of Conception Bay up to Holyrood,
where we anchored in Chapel Cove. The bottom
here being light-coloured, we could appreciate
the extraordinary clearness and transparency
of the water. Although we were close inshore,
it was eight or ten fathoms deep ; and yet we
could see the anchor lying on the bottom, and
the whole length of the chain up to the surface.
On the rocks nearer shore myriads of green
echini completely covered the bottom of the
water, except in patches where a kind of coral
forming crusts an inch thick, with small round
knobs rising from it, was to be seen : this coral
belongs to the genus myriapora, and is abundant on all the coasts of Newfoundland. This
afternoon I ascended the Butterpots, a remarkable hill immediately on the east side of the
inlet of Holyrood. We passed to the foot of
the hill through a good deal of oldb urnt wood,
now all white and brittle. The mosquitoes
were troublesome, but we kept them off in
some measure by smearing our faces with

camphorated spirits of wine and sweet oil, a
bottle of which I had provided for the purpose.
We unfortunately attacked the face of the
hill instead of working round it; accordingly
we had a very hard climb, and in one place I
slipped between two blocks of rock, and found
on arriving at the summit that I had thus
squeezed and broken the tube of the barometer. The summit is flat, with a precipitous
face towards the west, and a more gradual but
still rapid slope to the east : it probably attains
the height of 1000 feet above the sea. The
view from it was much the same as from the
Cats Cove Hills.

I did not see much of the people in this
part of Conception Bay, as I was too eagerly
engaged in hunting rocks, but as far as I can
recollect they seemed to be comfortable and
contented, and their gardens and small fields
to be superior in fertility to those in the neighbourhood of Harbour Grace or St. John's.

June 28th.—From Holyrood to Portugal
Cove, visiting Kelly's Island, Killegrews, and
Little Bell Isle on the way. From Holyrood
to Topsail there is a low tract of flat land,
about a mile wide, full of ponds, and having a

beach of large pebbles; and on the south-east
side of each of the three islands opposite to it
is a similar beach of a triangular shape, the
apex running out to sea and pointing to the
opposite shore. It looks as if a current swept
past the islands from the north-west, deposited a heap of pebbles in the slack-water
under the lee of each, and piled up the rest
of the pebbles on the opposite shore. As a
proof that there is such a current, several
things (and I believe the body of a boy) which
were lost from a vessel that was wrecked in
1840 near Pouche Cove were picked up in
Holyrood, having doubled Cape St. Francis
and travelled to the very head of Conception
Bay. When I landed on Kelly's Island I found
several workmen getting stone for the pro-
jected Catholic Cathedral in St. John's. The
island is composed of shale, with bands of hard
fine-grained gritstone : a thick bed of the
latter comes out in the cliff on the south-east
side of the island, and, as the shale beneath and
above it decomposes, it falls down and forms
a considerable talus of fragments at the cliff's
foot. One or two schooners were carrying
away these loose blocks for building-stones.

A few days before I visited them they had found at one place, beneath the roots of an old tree that grew over the rubbish, an old iron gun. It was a nine-pounder, very rusty, having no date or other mark than S. C. on each end of the trunnions, with a figure between the letters that might either have been a fleur-de-lys or an arrow-head. I was told afterwards that in some former wars the island had been used as a place of refuge for the inhabitants of the neighbourhood.

June 29th.—Having sent across to St. John's for letters, &c., we sailed in the evening to Carbonear, with a strong breeze of wind and a storm of rain.

June 30th.—To-day, being Sunday, the men, who were all Catholics, went to chapel, and I spent the day with some gentlemen who kindly invited me ashore. The land on which Carbonear stands is not so level as that of Harbour Grace, the town therefore is more irregular; and as several mercantile establishments which formerly existed there have been broken up, it has, in parts, a forlorn and deserted air. Nothing, indeed, looks so wretched as large unpainted and deserted wooden buildings.

I spent the evening at Mr. Packe's, who has a very pleasant place on the top of the ridge at the back of the town, where he has reduced several acres to a very good state of cultivation. I went over his farm, and saw oats just coming up, as well as a good crop of Timothy-grass, and some fine young cabbage-plants springing. These were reckoned early, and it was said in about three weeks the grass and the cabbages would be ready for cutting. Mr. Packe told me, however, that his farm, with all the advantage of his having many men who could employ their spare time on it, instead of yielding a profit, cost him nearly 100*l.* per annum to keep it in order.* I was much charmed with the aspect of a wild little valley behind the house, partly filled with wood and a succession of small lakes or ponds, through which a brook ran off towards the sea, while the view from the front of the house commanded the mouth of Carbonear Harbour and its island, and a large part of Conception

* I hope Mr. Packe will excuse my introducing him in *propria persona*, and reporting private conversation, but such facts as the above are valuable to show the nature of the country.

Bay, with Cape St. Francis and its lofty
shores.

July 1st.—Sailed with a light air along
shore to the northward, landing at several
points as I went along. About Western and
Northern Bays were several spots where the
slate rocks were covered with red ochre.
There appeared to be an oxide of iron dispersed through the rock, which, when it was
exposed on the surface, formed a peroxide.
It was traditionally reported that the Red
Indians used these places formerly, to collect
the red ochre which they smeared over their
persons, clothes, and instruments.

In the evening it came on to blow fresh
from the south-west, and we anchored in
Northern Bay. Went ashore, and caught
some trout near a pretty water-fall at the head
of the bay. A considerable number of houses
were scattered round the little bay, as about
all the small coves along this shore. The
people were busied about their gardens, as also
in spreading the cod-fish to dry on the flakes.
In both places they seemed to be greatly
annoyed and troubled by the mosquitoes. The
women, although the weather was very hot,

kept large shawls spread over their head and
shoulders, and pinned tight round their faces,
to protect them from the attacks of the flies
while engaged in their work. They were
very civil and hospitable, inviting me into
their houses, and offering spruce-beer or tea.
It was here I first noticed the swarms of small
fish called capelin. These fish (Salmo arc-
ticus) are rather longer than the hand, with
slight elegantly-shaped bodies, greenish backs,
and silver bellies, some of their scales being
tinged with red. They are very beautiful
little things, and in June and the early part of
July crowd into the shores in countless myriads to deposit their spawn. The head of
this little bay had a small strip of sand beach,
on which was a slight rolling surf, and every
heave of the wave, as it broke on the sand,
strewed its margin with hundreds of capelin,
leaping and glancing in the sun till the next
wave swept them off and deposited a fresh
multitude. The clear green of the water,
margined with a belt of white foam full of
these elegant creatures, glittering in the sun
shine, formed altogether a most beautiful and
interesting sight. We picked up a bucketful

of them as we were going off, for supper and
breakfast, as, when fresh, they are most delicate eating.

July 2nd.—Continued our voyage along
shore into Bay Verde, or, as the people call it,
Bay of Herbs. This latter name is singularly
inappropriate, as it is a wild desert place, composed entirely of bare red gritstone, like that
on the coast near St. John's. Bacalao Island,
which is close by, being almost inaccessible, is
the favourite resort of thousands of sea-birds,
which kept flying about us as we lay ˙ becalmed for some time in the bay. At noon a
fresh breeze from the south-west sprang up,
and we sailed through the passage between the
island and the main, and shaped our course
for Trinity Harbour, on the north-west side of
Trinity Bay. As we neared Trinity it got
very cold, the thermometer sinking to 48°, and
we found three small islands of ice a mile or
two to the northward of us.

July 3rd and 4th.—Trinity Harbour is one
of the finest in the world, having three spacious
basins entirely landlocked. A lofty rocky
promontory rises in the centre of the harbour,

on the lower parts of which stand the houses composing the town, scattered about with the usual irregularity. There is a good church and churchyard, and several large and excellent houses with pleasant gardens. The storehouses and yards of several mercantile establishments are also on a very extensive scale, though the trade of the place is, I believe, declining, and is greatly fallen off from its former state. The harbour is surrounded by rocky hills forming precipitous cliffs toward the sea, and no road has as yet been formed from it, though one to Catalina has been surveyed. The rocks about it are slate, with beds of greywacke conglomerate. The population of Trinity and the neighbourhood is about 3000.

July 5th.—Sailed in the afternoon. The three islands of ice off the mouth of the harbour were covered by gulls and other seabirds, that were continually wheeling round them. Several whales were spouting round us in the coves, close alongshore, apparently pursuing the capelin. Many of these whales were of large size, as was apparent from the

huge tails they heaved into the air. Put into
a little harbour on the south side of Buona-
venture Head.

July 6th.—After towing the vessel out of
the harbour, went away in the boat to examine
several islands. Saw a thrasher at a distance.
This is a large cetaceous animal, that lies on
the surface of the water, which he continually
lashes with what shows at a distance like a
great tail, but is, I believe, a large pectoral fin
or paddle. He makes the sea foam around
him, and is said to be then attacking the
whale. My men told me that, while the
thrasher thus persecuted the whale from above,
the sword-fish was thrusting him with his
jagged weapon from below, and one of them
assured me he had seen the whale rise to the
surface and heard him roar with pain at a
distance of three miles. As this was Kelly's
story, however, I consider the authority very
doubtful, to say the least of it. After rowing
all day in a burning sun among the islands,*

* From one island I saw two well-defined stripes of wind
blowing down the bay, each about one mile broad, with
three stripes of perfectly calm water alongside of them,
giving a very peculiar banded effect to the aspect of the

we saw the vessel still becalmed under the
land in the evening, and were obliged to pull
back to her, when a light breeze sprang up,
and we made for Smith's Sound.

July 7th.—Thick fog early in the morning,
at the mouth of the sound, but as we proceeded we found it not to extend far inland,
and we drew gradually out of it into a blazing
sunshine. The sail up the sound was very
beautiful. It is a fine river-like arm of the
sea, from one to two miles wide, with
lofty, and in places precipitous, rocky banks,
covered with wood. We occupied the whole
day in examining both sides of the sound, and
anchored in the evening at the head of it in a
still lake-like bay, at the north-west corner of
Random Island. There is here a shallow bar
connecting Random Island with the main
land, which is only passable at high water.
This we carefully examined to find the deepest
channel, in order to pass it in the morning.
The shore here, both on the island and the
main, is low and well wooded. It is com-

water of the bay. The wind seemed to blow for some time
in the same direction within the same narrow limits, but
gradually shifted its place and visited other parts of the bay.

posed of the same shale as Bell Isle in Conception Bay, and its soil is of a superior fertility to that of the rest of the country, and
timber is consequently better. We saw several
seals here, one of which we shot, but it sank
before we could get it. Although the nearest
shore was full half a mile from our anchorage,
we were immediately on our arrival invaded
by such hosts of mosquitoes, that, after vainly
trying to sleep for some time, we were obliged
to fasten down the hatches and skylights, and
then make fires of old rags, oakum, and other
things, till the vessel was saturated with
smoke.

July 8th.—Soon after daylight we tried to
tow the vessel over the bar, but found the tide
did not rise high enough to admit of it. It is
only at spring tides, or possibly during strong
easterly winds, that there is more than six feet
of water, while our vessel, the Beaufort, drew
six feet and a half. I accordingly put three
days' provisions into the boat, and, taking two
men with me, set off, appointing to meet the
Beaufort in Hickman's Harbour, on the south
side of Random Island. A fine mass of peaked
hills some miles to the west sent down some

spurs to the coast opposite Random Island, which I found to consist of sienite. The character of the scenery of Random Sound, on the south side of the island, is much the same as that of Smith's Sound on the north, wild and beautiful, and conveying, from its stillness and silence, the feeling of utter solitude and seclusion. As we rowed alongshore we came to a small rocky point of red sienite, on which stood two large masses of rock in a singular position, the one resting on the other, on the smallest possible base, apparently, on which such a block could be balanced. On this point was a she-otter with her young ones fishing, one of the latter of which I shot, as they were of good size. Here we dined, and presently after came to a large bay or recess to our right, from the head of which a well-marked valley ran away to the south. Here we got a breeze, and putting up our lug-sail we rattled away down to Hickman's Harbour. Landing at one place opposite this to make some observations on some rocks that caught my eye, as soon as we got under the lull of the woods we were assailed by mosquitoes. I endeavoured for some time to disregard them, but was at

last fairly maddened and driven off. They
came in such clouds before my face that I
could not see to write in my note-book, while
the eager voracity with which they fastened on
every inch of skin exposed was absolutely terri-
fying, so that at last we made a simultaneous
bolt into the boat and shoved off into the
breeze, which soon swept them away. We
then sailed to the mouth of the sound, but,
seeing nothing of the Beaufort, returned and
put into a little cove on the south side just as
it was growing dark. We were guided by the
sound of falling water to a little brook that
leaped over the rocks into the sea. Here we
hauled our boat ashore, and established ourselves on a little margin of shingle and broken
rocks between the woods and the water. We
gathered some old trunks of trees, and cut
down some fresh ones; fired some loose brown
paper out of the gun to light a match, and
made a fire, piled on some logs, and as soon as
they were fairly alight covered them with
wet moss to make a smoke to drive away the
mosquitoes, in which we succeeded pretty well.
The next thing was to boil the kettle, into
which we put a handful of tea and about

two table-spoonfuls of molasses, to make the usual beverage of the country. In the mean while a lot of green boughs had been cut to interpose between the rocks and our persons as a bed, and wood and moss piled for the night to keep up the fire and the smoke. The men had brought a spare stay-sail as a covering, and I had a cloak. We were obliged to tie our heads up in handkerchiefs, and put on gloves, to keep our hands and faces from being bitten; and, thus defended, hushed by the falling water of the brook and the splash of the sea three yards from our feet, we went to sleep. Although awoke once or twice by heavy rain hissing on the fire, I passed this my first night in the open air in tolerable comfort, but found in the morning a few sore places in my ribs from some sharp corners of rock coming against them.

July 9th.—On awaking found it high tide and the water nearly up to the fire. The tides on the coast of Newfoundland are generally so small, the water never rising or falling more than six feet, and the coast is usually so bold, the water being deep close in-shore, that they are practically disregarded, the only dif-

ference between high and low water being,
that a few feet more or less of perpendicular
rock are covered with water. The first thing
this morning I bathed, but, though the weather
had now been very warm for more than a
month, I found the water, even in this sheltered nook, bitterly cold, so that one plunge
from a rock quite sufficed me. We derived,
however, one advantage from it, as, while I was
dressing, my attention was caught by something moving on the bottom twelve or fifteen
feet below me, and I soon found it to be
covered with lobsters. One or two of these,
by means of a pointed stick, we managed to
capture. The singular clearness of the water
is most remarkable; when the surface is still,
the echini, shell-fish, and cretiniæ clinging to
the rocks, crabs and lobsters crawling on the
bottom, fish, medusæ, and myriads of sea-
creatures floating in its depths, were as clearly
visible to a depth of thirty or forty feet as in
air itself. Just as breakfast was ready it
began to rain again and the mosquitoes began
to thicken, so we pushed off in the boat, and,
hoisting an oar upon two forked sticks, we
threw the sail over it, and thus made a kind of

canopy, under which we breakfasted, without, at the same time, being breakfasted upon more than was endurable. It afterwards cleared, and we rowed about the shores of the sound, examining the cliffs and waiting for the Beaufort to make her appearance. The cliffs hereabout are of a dark-red thick-bedded hard sandstone or gritstone, and often rise from a considerable depth below the water to a perpendicular height of 300 or 400 feet above it, sometimes without a single break and scarcely a jutting ledge. Sometimes, however, a narrow talus is formed by huge fallen fragments, and on the older parts of these soil has gradually accumulated and trees have grown. Among the roots of these trees numerous otters have formed their burrows. These are called " otter rubs," from the smooth and beaten path leading to them, or from the polished stems and roots of the trees, against which the animals apparently rub themselves. As I sat in one of these places among the gnarled stems and twisted roots of some old trees, and looked up to the huge wall of rock on one side, and down on the other through the leaves to the boat waiting at the foot of the

rocks below, it seemed so admirable a subject
for a picture, that I more than ever regretted
my want of artistic skill to do justice to the
scene. As we sailed out of the sound a small
whale or grampus rose near us, and I fired a
charge of swan-shot into him, but he seemed
to treat the salute with the utmost indifference.
The seaward view was entirely obscured by
fog, which every now and then drove in clouds
up the sound, but soon dispersed over the land.
After landing in a small cove and boiling our
lobsters for dinner, we observed some fishing-
boats at Random Western Head, about four
miles off, and pulled across to them to know
whether they had seen anything of the Beaufort. While we were talking to them the fog
cleared off, and we saw her in the offing, standing in towards us. We then visited Long
Island and Heart's Ease, where there is a
small settlement of several houses, and got
aboard about dusk. After dark it fell calm,
and about ten o'clock I went in the boat to
find the entrance to a small harbour called
Fox Harbour. It was very dark under the
shadow of the cliffs, but we eventually hit
upon the spot, and found a channel between

two small islands. In returning I could not but be struck with the beauty of the scene. There was not a cloud in the heavens, and, though no moon was visible, the sky was flaming with stars; the planet Jupiter, more especially, cast quite a glare upon the still water, which being nearly landlocked might have been taken for a lake, but for the gentle heave of the ground-swell, which told of the stirring of the "ocean old." The air was mild, and yet the northern lights were glimmering with brilliant flickering streamers over the last faint reflection of sunset, while every dash of the oars and surge of the boat made the sea flame with phosphorescent sparks and eddying flashes of light. Around this brilliant scene dark headlands lay in stern yet calm repose. All was still and silent except the booming sound of a whale as he rose to blow, making a noise occasionally like the puff of a steam-engine, and the plash of the waters as they closed again round his descending carcase.

July 10th.—It having been quite calm all night, the vessel did not come to an anchor in Fox Harbour before six in the morning, and about eight I set off in the boat with four hands to examine the inlet called the south-western arm. We pulled along under the northern shore, visiting a fine-looking harbour called by the people Jones's Harbour. At the mouth of this a breeze sprang up, and we hoisted sail and ran alongshore. The men, having been up all night, fell asleep on the thwarts; and as I lay back between the yoke-ropes of the rudder, I could see no other moving thing on the broad sound, and the hills and woods around us, but our little boat dancing from wave to wave. Just before we arrived at the head of the arm dark clouds rose in the west, and a heavy thunder-storm came on. We caught sight, however, of a heap of chips of wood in a cove on the right hand, which the men said was probably near a winter-house or tilt. We made for it accordingly, and found a path leading in about fifty yards to a very good " tilt."

This was formed of trunks of trees placed upright on the ground close together, with larger ones for the corner-pieces, and a good strong gable-end roof formed of a frame of roughly squared beams. The corner pieces and beams were nailed together, and the rest

driven in tight with wooden wedges wherever
necessary. The interstices of those trunks
which formed the walls were filled up with moss
tightly rammed between them ; and the roof
was covered by long strips or sheets of birch
bark, laid tile-like over one another, and kept
down by poles or sticks laid across them. A
space for a door is left in the middle of one
side, and a fire-place is built up with stones
and boulders against one end, over which is
a space in the roof, and some boards nailed
together for a chimney. In this way a
tolerable room, twelve or fourteen feet by
eight or ten, is formed, sufficiently compact to
keep out wind and weather to a certain extent.
A " crew" of men, say six or eight, go off in
the beginning of winter with a stock of provisions to the head of some of the largest and
least-frequented arms of the sea they can find,
where the wood has been least cut up, and,
building one of these huts, employ themselves
either in cutting fire-wood to be hauled out
over the snow and ice, or in making oars or
staves, building punts, fishing-boats, or even
in some cases small schooners. The house
which we had thus lighted upon had been

occupied by a party the preceding winter
building a large boat, which was the reason of
the pile of chips we saw on the shore, being
the spot where they had built and launched
her. The roof had suffered a little, some of
the bark or "rinds" having been stripped off
by the winds. This, however, was soon re-
paired; and in a quarter of an hour we had
one end of the roof water-proof, and a blazing
fire lighted. We then brought the sails and
provisions out of the boat, and one of the men
"rinding" a few more trees, we completed the
roof, and commenced cooking our dinners.
The storm not seeming likely to abate when
the evening closed in, we determined to stay
where we were for the night. Accordingly,
one man took a gun to look for a duck, one
went to moor the boat more securely, and the
others made new hafts for their hatchets, while
I copied out my geological notes from the
field-book. Just before dark it cleared a little,
and I followed a narrow path into the woods.
After running through thick woods for some
distance, the path came out on a small marsh
at the foot of a rocky hill, and then divided
into two or three. I pursued each of these

to the end, and found them gradually to be
lost in the wood, usually ending at the
stump of some tree larger than common, and
which the men had gone farther in than usual
to procure. Beyond these points all was dense
and tangled thicket, now dripping wet. On
returning I found the boat's sail dried and
rigged into a hammock for me, and the spare
stay-sail hung round the fire at the back of a
heap of green boughs, forming a bed and
curtain for the men. We then supped on the
remains of the dinner, had a glass of grog and
a pipe, and spun yarns round the fire till we
all fell asleep. These men were all natives of
Newfoundland, but completely Irish, with all
the brogue, and some of the wit and humour, of
the latter country. They apparently delighted
in tales of the marvellous, which, if they did
not recollect, they were not slow to invent.
They were also evidently imbued with super-
stition. One of them, named Dick, was going
down to the boat soon after dark for something
he thought he had forgotten, when another
began a story of some man living in a lonely
place hearing a knock at his door one night,
which, on opening, he found to proceed from

Chap. II.] A GHOST STORY. 73

a headless visitor that walked in and usurped
his place at the fire. Dick, who had got to
the door, instantly returned and sat down, and,
on my asking him why he did not go, said,
" Oh, by God, sir, I'm feared to go after that !"
and to my great astonishment I found that,
though an active daring boatman, he was
absolutely afraid to venture fifty yards in the
dark, and that his fear did not seem unnatural
or ridiculous to the rest.

July 11th. — Breakfast at daylight, and
pulled back again along the opposite shore.
A flat valley ran from our resting-place northwards into Random Sound, and had apparently
formed at one time a water connection between
the two inlets, and thus rendered the land
between the sound and the south-west arm an
island. I had not time, however, to examine
it more in detail, as my business was to hunt
rocks, and if possible find coal, which I had
been told had been seen about Random Island.
From the middle of the south-west arm I saw
the same picturesque group of hills in the
country to the west which I had previously
observed from the sound. The head of the
arm was also composed of sienite, as was the

case with the sound, from which, to the sea,
the rocks were all clay slate, quartzose grits,
and dark slate rock. On going alongshore
we came to a small sand beach, on which were
the fresh tracks of a bear and a fox. They
both led into the wood, where it was hopeless
to attempt to find them. A little below, a
very large bird, a kind of brown eagle, with
whitish head and yellow claws, flew out of a
cliff, and flapped lazily along from cliff to cliff
before us. I sent a ball after him out of my
double barrel, but without success. At one
spot where we landed was a small brook with
some meadow-land covered with grass, and
spotted with blue iris and other flowers.
Contrasted with the general covering of dank
moss, it looked quite delightful. It did not,
however, run far, as 200 or 300 yards from
the shore the usual close wood and moss
covered the ground. Reached Fox Harbour
at three P.M., took in some water, got up the
anchor, and stood into the bay, bound farther
up for the Bay of Bull's Arm.

July 12th.—We beat up all day against a
south wind, and in the evening anchored in
Mosquito Cove, Bay of Bull's Arm. The aspect

of the coast and country was still the same,—
namely, wild and rugged, with scattered and
stunted wood. Beyond Fox Harbour and
Heart's Ease there were no inhabitants.

July 13th. — Not very well this morning,
but about 11 o'clock set off with all hands in
the boat for the head of the arm. Here we
landed near a little brook, where I left the
skipper Gaden and one man fishing, while I
with the rest started across for the head of
Placentia Bay, determined among other things
to procure some fresh meat of some sort if
possible. The neck of land here which con-
nects the province of Avalon with the main
part of the island is not more than three miles
wide, and, where we crossed, not above 150 or
200 feet above the level of the sea. There had
been some talk of cutting a canal through it
to connect Trinity and Placentia Bays, and
were those places the seat of a large mercantile population it would no doubt be effected.
At present, however, the only purpose it could
serve would be the occasional passage of a
fishing-boat from one bay to the other, according to the season and the relative abundance
of fish. There are still the marks of a path,

E 2

along which, according to the tradition of the
country, the French, when they were in possession of Placentia, used to haul their boats
from one bay to the other on the dry land.
The slope towards Trinity Bay is gradual, and
over marshy · ground, on which we found a
great many tracks of deer apparently freshly
made. On the Placentia side the descent is
more abrupt and rocky, and through a thick
wood. On the barren, at the top of the ridge,
I shot a brace of ptarmigan without respect to
the season of the year, a consideration but
little attended to by a hungry man in a wild
country. At the foot of the woody descent we
came to a pretty brook with a depth of water
of two or three feet and a rapid current. Its
bed was, however, rocky, and it shortly spread
out among the pebbles and sands at the head
of the bay we had come out upon, and which
went by the name of " Come-by-Chance."
Walking about a mile down the brook we
came to a lone house at its mouth, with six or
eight cows feeding on the meadows, formed by
the washings of the brook and the sea. It is
only in such situations, where the alluvium of
a brook has had room to spread itself out, that

natural grass is to be seen in the country. We
found only the woman and children at home,
the man having gone down to fish about Placentia. They managed the cows and the
salmon-nets. One of the latter was placed at
the mouth of the brook, and one a mile or two
up ; and they told me that last year they had
caught six tierce of salmon. These varied in
value from 3*l.* to 6*l.* currency, so that the wife
was enabled to get 20*l.* or 30*l.* every summer
by her own exertions, independent of the
results of her husband's cod-fishing, which
would probably be as much more. As they,
of course, paid no rent for either house or land,
and had wood for firing and boat-building at
command, they were well off. To counter-
balance these advantages, however, they had
no fresh meat unless when the man shot a
deer, and had to fetch their bread, salt pork,
tea, sugar, and every article both of clothing
and food, from Placentia, or some other distant
merchant's store, where they necessarily paid
high prices from the want of competition and
the nature of the trade. We got here some
fresh milk and butter, and four salmon, and
feasted accordingly, and while we were strolling

about, the husband's boat came "goose-
winged" up the bay before a fresh breeze, and
my men got what they enjoyed more than
even salmon and milk, namely, some news
from Placentia and St. John's. I also heard
of two gentlemen of St. John's who lost their
way in the country a few days before, in
attempting to walk across this neck of land a
few miles farther down, where it is about nine
miles wide. They were out in the woods two
days and a night in dreadful weather, with no
shelter and but little food, and with difficulty
got back to the place they had started from.
They were accustomed to traverse the country,
and not easily disheartened, so some notion of
the difficulty of penetrating it may be gained
from this fact. We walked back to our boat
in the evening, and found Gaden and the
other man had caught no fish, but had been
nearly eaten alive by the mosquitoes.

July 14th.—A dense fog this morning, which
about ten cleared off, and I took the men and
the boat, intending to visit a lofty hill called in
Bullock's chart " Centre Hill," between Bay
of Bull's Arm and Deer Harbour. In climbing up the first bank I found myself very

weak and unwell, with sickness and frequent
giddiness, and it was with much difficulty I
toiled over the marshes and through the tangled woods to some craggy and broken rocks
with dark pools of water at the foot of the hill,
which looked very steep and inaccessible.
After taking a little biscuit soaked in whisky
and water, however, and a strong pull at the
" creature itself," I managed to reach the top.

The view was very extensive, commanding
the greater part of Trinity and Placentia
Bays, the high lands about the Bays of Bonavista and Fortune, and a wide tract of the
interior. The interior consisted almost entirely of marshes, dotted with ponds and
patches and strips of wood; a few ridges
showed either bare rock or white lichens to
contrast with the velvety green and yellow of
the marshes, which are altogether deep
wet moss. Although the sea was within five
miles of us on one side, and the woods and
hollows obscured them at a greater distance
than eight or ten miles on the other, I counted
no fewer than 152 ponds from the summit of
this hill, none less than twenty or thirty yards
across, and some of them half a mile or a mile

wide. The height of the hill was probably
1000 or 1200 feet. We could trace several
well-beaten deer-paths traversing the marshes,
but saw no deer. I found, however, a bag of
buck-shot on the top of the hill, which is used
as a look-out by persons who come to shoot
deer in the fall of the year. In descending we
found a kind of track that led us out by a
shorter way than the one we came, and close
by the shore we found a kind of rude sledge
that had apparently been hastily constructed
to haul out the body of some deer over the
snow. On returning to the vessel found myself very unwell, but was cured by a boiled
ptarmigan and a basin of the broth. An
English stomach accustomed to fresh meat is
really very inconvenient to a traveller. In this
hot weather I had loathed the salt pork and
beef, and lived principally on tea and biscuit.
Our water, too, lately had been got principally
from brooks that flowed from the marshes, and
was full of vegetable matter ; so that, from bad
living and the incessant attacks of the mosquitoes, I was nearly driven into a fever, and
could neither eat, sleep, nor think.

July 15th.—Felt much better. Weighed

anchor and sailed along the shore towards
Tickle Harbour. In the middle of the day
we got becalmed, and a thick fog came on,
and presently after a deluge of rain, that soon
poured through the skylight and ceiling into
my cabin, wetting everything except my bed.
In the evening it began to blow, and we met
some fishing-boats that told us Tickle Harbour was a bad place to lie in, and one boat
piloted us into a small cove called Chance
Cove a mile or two astern, where we arrived
just as it got dark, through a narrow and
rocky entrance. The man directed us, if it
came on to blow from the north-east, to get a
line ashore and haul behind a small rocky headland. After a glass of grog he sailed away
into the mist to go home, there being some
houses in a small cove a mile or two off towards Tickle Harbour.

July 16th. — Fine morning and pleasant
breeze, and we sailed alongshore round Tickle
Harbour Point into Collier's Bay, landing occasionally of course to examine the rocks. In
the evening one of the men, in getting water
from a brook, had his eyes nearly closed up
from the swelling caused by the bites of a very

minute insect called a sandfly, one of the three
kinds of biting flies which go under the general denomination of mosquitoes.

July 17th.—Saw some large flocks of ducks,
which were, I believe, the eider-duck, but could
not get a shot at them. We then visited
Chapel Arm, round which are some very picturesque peaked hills, clothed with dense
wood. Here, at a place called Norman's Cove,
we saw a small schooner on the stocks, and
found several houses. Sailing by a fishing-
boat we asked them for some fish, when they
threw half a dozen fine cod aboard, without
thinking of asking a recompence, or staying
to receive our thanks. Coasted along and
anchored in the evening in New Harbour, a
shoal place and difficult of entrance. We here
found several merchants' stores, and a considerable population, the beginning of a road,
and a bridge over the brook. I also got some
provincial newpapers, and found a paragraph
in one in which I was said to have discovered
copper, gypsum, coal, limestone, and, if I recollect rightly, silver, the account, I suppose,
of some imaginative fisherman.

July 18th.—Mr. Nieuhook of New Harbour

treated us very kindly, and piloted us out in the morning, and we sailed with a light breeze down to Heart's Content, along a low shore of bright red and grey slate. The wood along this was finer than ordinary, and the land apparently more fertile, grass growing occasionally, and in one place, called Green Bay, I observed some white clover in flower. Heart's Content is a fine spacious harbour, nearly circular, and excellently sheltered from all winds. The land about is high, and from this place the coast is lofty and bold for some miles towards the north.

The Reverend Mr. Hamilton came on board and invited me ashore. He has charge of all the eastern shore of Trinity Bay, as Mr. Bullock had of Trinity Harbour and the western shore, Heart's Content and Trinity being the respective head-quarters. Almost all the eastern shore of Trinity Bay is inhabited, and Heart's Content contains two or three mercantile establishments or agencies, and a considerable number of houses. The different places alongshore are only connected by miserable winding wood-paths or narrow boggy tracks over the roots of trees, but Heart's Content has

a road, surveyed to Carbonear, which is completed except three or four miles of " barren "
in the middle, and all the brooks are bridged
over.   There is also a road marked out from
New Harbour to Spaniard's Bay, and wooden
bridges constructed over the brooks.

We were detained a day at Heart's Content
by bad weather and a gale from the north-east,
and on July 20th sailed again alongshore,
calling at New and Old Perlican, Hants Harbour, &c., down to Baccalieu Island.   Here
the wind fell to a calm, and it was morning of
July 21st before I landed at Quiddy Viddy,
and walked up to St. John's, after a trip of five
weeks.

CHAPTER III.

Leave St. John's for the Westward—Lamelin—St. Pierre—
Langley—Bay of Islands—Excursion up the Humber
River—Calm in the Gulf of St. Lawrence—Catch of
Cod-fish.

NOT having succeeded in finding coal or other
valuable mineral matters in this part of the
island, the governor thought it would be ad-
visable for me to go round to the western side
to examine the coal which was creditably reported to exist there, and then work my way
home along the south shore of the island, and
examine the other side of Avalon in the fall of
the year. We accordingly attempted to re-
place one of our men who was dismissed for
drunkenness by some one who had a knowledge of the coast we were going to visit, and
who could act as a pilot: we were, however,
unsuccessful in this, but engaged an active
young fellow to serve as a common sailor, and

determined to trust for guidance to our charts and our own skill or good fortune.

July 30th. — After the usual trouble in getting all our things on board which attends the departure of both large and small vessels, we sailed to-day at twelve o'clock with a fine breeze from N.W., and passed Cape Broyle about six in the evening, with the wind lulling.

July 31st. — Having had a light wind all night, we only doubled Cape Race early this morning, and about eight found ourselves off Trepassée, among a fleet of fishing-boats, all at anchor, and all catching fish with great expedition, and sailed with a light easterly wind for the island of St. Pierre. The practice I had already had did not prevent my feeling very sea-sick, which compelled me to take refuge in my berth the greater part of the time.

August 1st.—Dense fog, so that we could not see twenty yards in any direction. About eleven o'clock it cleared off a little and we saw land ahead, which turned out to be Laun Islands, the current having drifted us in-shore more than we expected. The fog closed in again, and the wind freshened, with

a short jerking sea, and about six in the evening we found ourselves close in under some
high barren shore, round which the fog was
whirling in white eddies, utterly obscuring its
features. We concluded this to be St. Pierre,
but as none of us had ever been there before,
we did not venture to try for the harbour,
which is narrow and rocky, and so stood off
again. Another attempt just before dark was
equally unsuccessful, so we stood off for the
night on a S.S.E. course, with a fresh breeze
from south-west and a rough sea. At mid-
night, when we calculated we had made fifteen
or twenty miles offing, we hove to.

August 2nd.—Stood in for the land again,
the fog being as dense as ever. We very
shortly came upon a fishing-boat at anchor,
and found that the strong current setting into
Fortune Bay, between St. Pierre and the main,
had drifted us close in-shore upon a highly
dangerous coast, whence reefs of rock extend
in some places three miles out to sea. Luckily
we had hit a spot where we had the little harbour of Lamelin under our lee, and by following the directions of the fishermen, and leaving
one rock on the starboard, another on our lar-

board hand, and then hauling up to windward, we found ourselves in smooth water, where we anchored. Still the fog was as dense as ever, and what kind of a place we were in we had no notion. We heard, however, frequent voices, and presently, as it began to clear a little, we observed on one side of us a forest of masts looming through the fog, apparently large vessels, but which turned out to be a fleet of fishing-boats waiting to put to sea. Then the dim form of land was seen, and when the fog dissipated we found ourselves in a snug berth enough, behind two islands, with a flat marshy shore on the other side, rising into low barren rocky hills in the interior. We were detained here till the afternoon of August 4th by strong westerly winds, when, growing impatient, I put to sea to try to beat up to St. Pierre. There were a considerable number of houses in Lamelin, but, from the neighbouring country being destitute of wood, most of the inhabitants shift their quarters, when winter comes on, to the shores of Fortune Bay. Two men living here, however, had about fifty head of cattle apiece, which they kept on the adjacent marshes, where a rank grass grew.

There was a shallow salt lake at the back of
the harbour that filled at the rise of the tide,
and was called by the people a Barrasway.
This is a very common term for a shallow
marshy inlet or salt lake along the south
coast of Newfoundland. It is spelt in the
French charts Barachois, and is, I conclude,
a Norman word. A Nova Scotian schooner
came in while we were at Lamelin, "The
Betsy of Halifax." She was loaded with
pork, molasses, rum, and shop-goods, the
latter designation including all kinds of
printed calicoes, cloths, ribbons, gloves, shoes,
&c. &c. She traded along the coast, receiving
dry fish in exchange, and the owner of her,
while talking with me, pointed out a common-
looking fisherman, who, he said, owed him
100*l.*, and whom he immediately proceeded
to dun accordingly. I was surprised to find
them dealing in so large a way, and trusting
any of the tenants of the miserable-looking
wooden huts around us to such an amount.
The resident merchants complain dreadfully of
these floating shopkeepers, whom they consider as interlopers, and who often, by sudden
temptation of rum or finery, succeed in getting

off a whole cargo of fish that was justly due from the fisherman to his merchant for goods received. The ultimate result of the practice, however, may be good for both parties, as I shall have occasion to observe when speaking of the trade of the country. Meanwhile, the owner of the " Betsy " gave the people of the neighbourhood by no means the best of characters in one respect, as he said one winter he was utterly wrecked on Point des Galles, a few miles to the eastward, and while walking nearly naked on the snow, to try and recover a few articles of clothing which were washed ashore from the wreck, two or three men came down, and made no scruple to pilfer the things before his eyes, and would hardly give them up when demanded.

August 5th. — After buffeting about for nearly thirty hours, we managed to get into St. Pierre outer harbour just before dark. A subordinate officer came on board from the guard-ship (which was only a schooner of four guns) as we stood in, and pointed us out a berth to anchor, and requested to know our business. I gave him my card, and in a short time he returned with the commander of the

guard-ship, who was also captain of the port.
He behaved very civilly, and promised to
report my arrival to the commandant, and
demand permission for me to wait upon him.
Next morning, before I was up, a midshipman
came on board with the commandant's desire
to see me at twelve o'clock, when I accordingly
called upon him, and found him a fine intelli-
gent old man, who had formerly been an aid-
de-camp to Marshal Ney.   He was very civil,
and invited me to dinner.   I learnt here that
the Cleopatra, Captain Lushington, the frigate
on the Newfoundland station this summer, had
sailed the day before we came in, which was
a disappointment, as I had hoped to have met
with her, and formed an acquaintance with
her commander and officers.   The islands of
St. Pierre, Miquelon, and Langley are the
only territorial possessions left to the French
in this part of the world.   The harbour of
St. Pierre consists of an outer road, which is
protected by several small islands and rocks
from most winds; and an inner harbour,
which is smaller, and has a rocky bar that
does not allow of the entrance of anything
larger than a brig of about 200 tons.   On

one side of this inner harbour is the town,
consisting of a small narrow street of wooden
houses, few of which are two stories high, and
several lanes of still smaller dimensions. The
house of the commandant is tolerably sized,
built of wood, with several comfortable apart-
ments, having a small esplanade, and a couple
of guns, and a sentry or two before it. By
treaty the French are not allowed to erect for-
tifications, nor to have more than fifty soldiers
on the island at one time. There were two or
three large brigs and a ship in the outer roads,
and several smaller vessels, brigs and schooners,
in the inner harbour, besides many large boats.
They have very strict regulations in the port.
No English boats or vessels are allowed to
come in having fish on board, on penalty of
being seized; and no Englishman is allowed
to bring English goods and manufactures, or
to set up a shop in the town. There is, however, an American warehouse belonging to
Atherton and Thorne, which seemed to be
doing a large business. The island of St.
Pierre consists principally of sienite, and is
barren in the extreme. It is a mass of rocky
hummocks, the hills rising to a height of 400

or 500 feet directly from the water. The hollows and flatter parts consist of marshes and ponds; and there is not a tree in the island six feet high, a few scrubby fir-bushes alone contriving to exist. It is with the greatest difficulty that they have scraped together near the town sufficient soil to grow a few cabbages. They are entirely dependent on Langley and Miquelon, or the English settlements in Placentia and Fortune Bay, for firewood. Vegetables they get from Boston or Prince Edward's Island.

August 7th.—Set out early this morning in the boat for Langley, where the commandant told me he had a summer residence of a much pleasanter character than the one at St. Pierre, the country about being finer and more fertile. Just north of St. Pierre, and separated from it by a narrow channel is a small lofty island called Colombier. Its resemblance to a dove-cote arises from the multitudes of puffins which breed there, and are always flying about it in great flocks. From this we had a hard pull of three miles in the teeth of the wind to the cliffs of Langley, as the Little Miquelon of the French is called by the English. We then rowed round Cape Percée, which derives its name from a singular arched rock, and entered a wide bay that lies between Langley and the greater Miquelon. The scenery here was very striking and picturesque, as the high land of Langley sloped down towards the west, covered with rich green moss and skirts of wood, into a dense mass of wood of a finer appearance than usual, and covering some flat land that swept round towards the north, exposing a fine sand beach, and connecting the two islands of Great and Little Miquelon. These islands, not more than sixty or seventy years ago, were quite distinct, and are marked so on all old charts ; a considerable channel of two fathoms' depth running between them. This, however, is now entirely filled up, and a long narrow line of sand-hills with a beach on each side occupies its place. Instances have been known, even of late years, of vessels in stress of weather making for this channel, and being wrecked on the sands. On walking out along the western side of this neck of land, we found the remains of three wrecks there then, as the place with a southwest wind is very dangerous, more especially in foggy weather. I quite rejoiced in a walk on this sand-beach, the only one I had yet seen about Newfoundland, and indeed the only place I had yet found (except the roads about St. John's) where it was possible to walk at all, as one is accustomed to do, with the head up and the eyes off the ground, and without the constant fear of coming down on the nose. It was quite refreshing to be able to step out, with a fair " toe and heel" step, instead of scrambling among bushes, or floundering in wet marsh. The immediate neighbour of this strip of sand, however, was a large marsh, in which I shot half a dozen snipe as a compensation for its toilsomeness; and some part of it had been drained and was converted into meadows, on which were sheep and cattle. There are still more extensive meadows, I believe, farther north, on the Miquelon side, where enough sheep and cows are fed to supply St. Pierre and the neighbouring population.

The commandant's house is on a gentle elevation on the south side of the bay near a rocky brook, and prettily situated. Close by was a smaller house occupied by two gendarmes, who had care of the premises. One

of the gendarmes, named Ducroix, was married to a pretty young Irishwoman, whose
history was rather remarkable. She had left
Ireland with her father and uncle when the
cholera was raging in it, and had gone with
them to Quebec. On arriving at Quebec they
found the cholera was very prevalent there
also, and they went on to Montreal, where it
was worse. At Montreal both her father and
his brother were attacked and died of cholera,
and she was left alone in a strange land. She
managed to get back to Quebec, and there
procured a passage in a vessel going to Ireland,
intending to return to her friends; but the
vessel was totally wrecked on the beach of
Langley one night, and the crew and passengers with great difficulty saved. Here, being
very young and quite destitute, she was per-
suaded to stop and marry the gendarme,
who happened to be present, and to have assisted in rescuing her. This was six years
ago, and she had now four children. His
period of service was nearly out, and he expected to be ordered home the next autumn,
when she hoped to be able to visit her friends.
They behaved very hospitably to us, and gave

us a loaf of bread and a lot of lettuces, no contemptible present to a sea-voyager.

August 8th.—Having profited by our visit to a French port by taking in a sea stock of wine and brandy, we sailed from St. Pierre in the middle of the day with a south-east wind and thick and rainy weather. We could just make out the dim form of Cape Miquelon as we rounded the north end of the island, and then steered away on a west-north-west course into the fog.

August 9th.—Dense fog, with occasional squalls of wind and rain. I experienced a peculiar feeling of loneliness to-day, in looking out from the deck of our little vessel into the heavy fog resting on the dark boiling waves, as we slowly toiled over their crests, or sunk into their hollows, that I do not recollect to have felt before. We seemed so utterly shut out from all the world, so hidden by the dense curtain about us, and our vessel to be such a mere nut-shell on the wide waste of waters, which, as its boundaries were no longer visible, appeared to the imagination more vast than when defined by a sensible horizon ; and we passed so many hours and traversed so

many miles without change of any sort, in the
same damp fog, over the same dreary water,
that the fancy became affected with a kind of
awe, very different from the usual excitement
of a mere coasting voyage. I do not know
whether the rough minds of my companions
were affected in the same way, but they were
all silent also.

August 10th.—About six this morning,
through a slight opening in the fog, we saw the
dim outline of land a mile or two on our starboard hand, and accordingly hauled off to the
west-south-west. About noon the fog suddenly cleared off, or rather we ran out of it,
and found ourselves fairly in the Gulf of St.
Lawrence, with Cape Ray about ten miles
east-north-east of us, and a bright sun overhead. We then steered north for the Bay
of Islands, and at night were off the mouth of
St. George's Bay. It was a beautiful evening,
with a fine aurora in the northern sky, consisting of a diffused yellow light with brilliant
streamers rising in the north-west, the whole
dabbled with small black clouds.

August 11th.—Fine morning, with fresh
breeze from the south. Kept close in to Cape

St. George, and between it and Red Island.
Red Island is composed of horizontal beds of
red sandstone, while the cliffs of Cape St.
George are a light yellow limestone, chiefly
magnesian. We then passed along a low and
level shore, being a neck of land running on
one side of a large bay, called Port au Port.
This was covered with wood. North of the
entrance of Port au Port the coast is very bold
and lofty for a considerable distance, as far
north as Cape St. Gregory at least, the high
lands of which were visible as we came to the
entrance of the Bay of Islands. We were
here entirely in a new country, but the chart
showed us two small harbours on the south
side of the bay, called York and Lark Harbours, in the former of which we anchored
about dusk. We had some difficulty in getting
in, as under the high lands round the harbour
we were every now and then totally becalmed;
then a squall, rushing down some narrow
ravine, would suddenly strike us, and hurry
us along, "scuppers under," straining the
masts and "making everything crack again."

August 12th.—At sailing this morning we
kept under the southern shore, passing by

F 2

several low flat islands on our larboard hand,
in order to enter the most southern of the
three branches into which the bay divides,
and which is called Humber Sound. I had
been informed that at the head of this was the
mouth of the largest and most navigable river
in the island : this I was desirous to ascend as
far as possible, in order to get some notion of
the interior of the country. In entering the
sound a lofty ridge of high land rose immediately to the south of us, called the Blow-me-
down Hills. Its general elevation must, I
think, have been upwards of 800 feet. It was
bare at top ; and in some hollows, just under
the top and sheltered from the south and the
rays of the sun, some patches of snow twenty
or thirty yards across still remained, although
the summer hitherto had been very hot. The
slope of these hills, and the flat land at their
foot, was covered with dense wood. Where we
lay last night there was no sign of habitation ;
accordingly we were yet ignorant whether we
should find any people in the neighbourhood,
or, if we did, what nation they would belong
to. Just at the mouth of the sound, however,
we saw a small hut, and made towards it, and

presently a human figure presented itself in a
tattered dress, but it was not till we heard
him speak that we knew whether he was an
Indian or a European. We found him to be
an Englishman, who, with his wife and two
children, had just settled there, and that this
hut was his summer residence, his winter
house being back in the woods. He was in a
poor state at present, but expected to be more
comfortable in a few years. He informed us
that there were four other small settlements in
the sound; accordingly in sailing up we saw
on each side a patch or two of garden-ground,
with a house attached, forming a green opening among the black woods. The sound is
about a mile wide and seventeen long, and
the scenery, as might be expected from the
rocky and woody character of its shores, is of
a pleasing kind. We anchored at the head of
the sound in not more than eight feet of water,
with a muddy bottom. We found here an old
man residing with his family in a small
wooden house, with a garden attached. He
had lived in this spot for sixty years, and had
seven sons; one of these, a cripple, was with
him; the other six he said were away in the

woods hunting either for deer, or beavers, otters, martens, and other fur-bearing animals.

Our first object was to visit the river, which we rowed up about a mile. It had a width of fifty or sixty yards, and a depth of several feet, with a tolerably gentle current. A little farther up, however, the men at the neighbouring hut assured us there was a long rapid, and that our only means of ascent would be hauling the boat by a long line. On returning to the vessel we set to work to prepare, packed up four or five days' provisions, got ready the lines, got out the square sail for a tent, &c.

August 13th.—At seven this morning we set off. After rowing about a mile between wooded banks we came to a part where the river made several short turns through rocky precipices of white limestone and quartz, frequently eating its way into their bases, and leaving deep overhanging shelves. At about three miles we arrived at the foot of the rapids, where the scenery was highly striking and picturesque, lofty cliffs of pure white limestone rising abruptly out of the woods to a height of 300 or 400 feet, and being themselves clothed with thick wood round

their sides and over their summits. Three
men now took a long towing-line, while the
skipper and another man stood at the bow
and stern of the boat to guide her in the
eddies and among the rocks and boulders.
There was but little strand for the men to
walk, and they had occasionally to wade
through the water, or clamber along the edge
of the woods, passing the rope from one to
the other round the trees. However, after
some hard work, we at length reached the
head of the rapid, which is nearly a mile long,
and stopped to rest and refresh on a bank
of sand above it. Several seals rose in the
still water above the rapids, but took care to
keep out of gun-shot. One or two passed
close by us in the rapids, but we were then
too busily employed to think of shooting them.
Above the rapids the river has a more straight
course, and the hills recede from it without
losing their height, enclosing a valley about
two miles wide. This valley is filled with
groves of birch, and several small brooks fall
into the river. The river itself widens to 100
or 150 yards, with a shallower bed, sandy
shoals existing in places, with only narrow

channels sufficient for our boat between them.
One straight reach, however, is two or three
miles long and five or six feet deep. Above
this, and about six miles above the lower
rapids, we came upon some others. The
upper part of these rapids, for about a quarter
of a mile in length, is much more difficult
and dangerous than any part of the lower
rapid. The principal mass of water rushes
with the greatest force and rapidity over large
boulders, and sweeps against a steep rocky
bank. The bank affords no footing, and the
force of the water would not admit of any
boat being towed up it. We accordingly
tried the other side of the river, but here there
was no continuous channel of sufficient depth.
We were therefore obliged to unload the boat,
carry the things above the rapids, and afterwards fairly lift the boat over the rocks from
one little pool to another, and drag her in the
best manner we could. With much labour we
got her at last into deep water, and again proceeded with the oars. About half a mile
above these upper rapids we suddenly came
out upon a lake. This was a very beautiful
sheet of water, two or three miles across, and

with no land visible at its farther extremity, which bore from us about north-east. Some strips of sand formed its banks just at the entrance from the river, on which were several fresh deer-tracks. The sand, however, soon gave place to boulders, upon which it was scarcely possible to walk, from their round, smooth, and slippery surfaces. This border of boulders was about three or four yards across, with a steep slope, and above it was the usual bush of fir-trees growing out of piles of soft moss. As the evening was now closing in, and it began to rain, we hauled our boat up on the southern shore of the lake, and selected the smoothest place we could find among the wood-covered rocks for our bivouac. We cleared a small space by cutting down a tree or two, rigged one of these across the branches of two that were left standing, threw the sail over it, stretching which out towards the ground we pegged it down, and thus made a very sufficient tent, which, though it rained hard during the night, kept us tolerably dry. The dampness of the ground we partly avoided by lying on the fir-boughs, the trimmings of the trees we cut down for fire-wood.

August 14th.—I was awoke at daybreak
this morning by the cry of the "Loo," or great
northern diver, a very handsome dark bird
with white spots, and almost as large as a
goose. Its cry is a wild unearthly yell, with
a rather musical cadence, and sometimes a
sharp termination. It might be imitated by
sounding with a shrill and prolonged note the
words "yă hōō," and ending by a short
"chuck." From the loudness and closeness of
the sound this morning, I concluded one to be
close alongshore, and stole quietly down with
my gun, but could only see two at the distance
of at least half a mile. The water was perfectly still, and under such circumstances it is
astonishing to what a distance their cry will
be heard along its surface. It was a dark
heavy morning, but as soon as we had breakfasted we proceeded along the shore of the
lake, shooting a young gull by the way, to a
small brook coming out of a narrow valley at
the south-east corner of the lake. Here I shot
a couple of beautiful grey-spotted kingfishers,
which were new to my eyes, but are, I believe,
common in North America. A little breeze
soon after sprang up, and the clouds dispersed,

and we sailed with a fair wind up the lake. The high ground around the south-west end of the lake gradually slopes down into a flat country at its north-east extremity, the hills being close to the lake on its south-east side, but on its north-west side gradually receding from it and running off in a connected chain of rugged hills as far to the north as the eye could follow them. When we had sailed about half way up the lake we could just discern the tops of the low woods in the flat land round its northern extremity, from which circumstance, and from the time we took to traverse it, I judged it to be about fifteen miles long.

At its north-west corner we found the river again, coming in, in two branches, each about fifty yards across and several feet deep; these branches joined in about 200 yards, enclosing thus a delta whose base was about 200 or 300 yards in length. The banks of the lake hereabouts were flat and sandy, with occasional small marshes. Proceeding up the river, we found its banks, rarely rising more than ten or twenty feet, everywhere covered with dense wood, consisting of fir, larch, spruce, birch, and pine, many trees being of good size, and capable of affording good timber. The river was frequently 100 yards broad, but got more shallow as we proceeded, and at one or two places we had some difficulty in finding water enough for our boat. After proceeding about five miles against a tolerably rapid stream, we encamped on the north side of the river among a grove of birches. We had seen but little game, consisting only of two geese and a few divers, and had lost much time in attempts to procure these, having only succeeded in adding two or three divers and the young gull to our stock of provisions. These, with the addition of a lump of salt pork, we boiled all together in our boat's kettle, and, thickening the broth with a little flour, we generally cleared off all its contents, both fluid and solid. To the bouillée of the kettle we added some molasses-tea and a couple of common sea-biscuits both morning and evening, taking a biscuit or so during the day as we had leisure or appetite.

August 15th.—On coming out of oūr tent this morning I could hear, the air being perfectly still, a dull but continued noise of falling

water, which promised ill for the farther extension of our journey. However, we set off
in a thin mist, and after getting a mile or two
farther came to some islands on the north-west
side of the river, opposite which another river
came in from the north-east. Among these
islands we saw several wild geese, but the
noise of our oars had rendered them too alert
to be approached within gunshot. Just above
these islands the mist suddenly cleared off, and
we saw some great rapids right a-head, whose
foaming waters seemed to offer an insuperable
barrier to our farther progress. Nevertheless,
we landed at their foot, and, taking one of the
men, I set off to see how far they extended.
After a most difficult and toilsome scramble
through the woods, which were rendered
worse even than usual by a quantity of low
alder-like bushes growing on the banks of the
river, we came out in about half a mile on the
head of the rapids, and found a deep still
reach stretching off some distance beyond.
Eagerly desiring to penetrate along these unknown waters, I sent the man back with directions to bring up the boat, and I would wait
for them. After waiting an hour or two with-

out seeing any signs of it, however, I retraced
my steps and found the party comfortably
seated round a fire with the tea-kettle boiling,
and on remonstrating with them the skipper,
Gaden, said it would be impossible to drag
the heavy punt up these rapids, and, moreover,
that they had trusted to killing more game,
and had not more than a day's provision left.
The rapids were certainly worse than the last
and longer, and the boat was better fitted for a
heavy sea than a shallow river: still I thought
it possible to get it up; but I saw the men
were gloomy, and did not like the job, and to
do it would require them to "work with a will."
The most serious obstacle, however, was the
want of provisions, as, if we knocked a hole in
the boat against one of the rocks in the rapids,
and were thus delayed any time, we might
be a day or two without food. I accordingly
reluctantly consented to go back, under the
delusive hope of being one day able to return
better provided to overcome the obstacles of
the route. I insisted, however, on exploring
the other river which we had passed, and
which came in from the north-east, and, if that
offered an easier passage, ascending it as high

as possible, and trusting to our guns for support. This other river made several short
turns and then became wide and shallow, and
we came shortly in sight of a set of rapids
quite as bad as those we had just left. The
banks of the river here were open and pleasant, and on one sandy point we found the
skeleton of an Indian wigwam, consisting of a
circular set of poles sloping inward and meeting at top, forming a rude cone. We now
turned about, and sailed rapidly down the
river into the lake, on reaching which we
encamped close to the mouth of the left branch
of the river. The view of the lake, here
receding from the flat country into the bosom
of the hills round its south-west end, was very
beautiful, and superior to anything I had yet
seen in the country, the features of both
ground and water being broader and larger,
and not frittered away into mere picturesque
nooks and corners, as is usually the case in
Newfoundland. The hills at the south-west
end of the lake are gneiss and mica slate, containing quartz rock and primary limestone.
The flat country is composed of brown and
red sandstones and conglomerates. We found

this evening near our tent a trap for foxes and martens; it consisted of the upright stump of a tree partially hollowed out, and four boards nailed over it, forming a box, open on one side; a heavy log of wood fell across this side on the withdrawal of a slight prop connected with a trigger inside the box; on this trigger was a piece of venison for a bait, and when the animal seized it it brought down the log of wood, which broke its neck. It belonged, we afterwards discovered, to the man living down at the mouth of the river.

August 16th.—There was heavy rain last night, but our tent was waterproof. To-day we pulled down the lake against a brisk wind, and dined on a little island at the bottom of it, on which were many bilberries, or whortleberries, which the men called "hirts." We then went to the rapids, and with some difficulty "eased" the boat down by a line in the best channel we could find. On arriving at the lower rapids, however, we determined on shooting them, which we did safely, and, after again admiring the striking scenery of the cliffs and woods around them, we reached our vessel at sunset.

August 17th.—I had been too eager to set off, at starting, to extract information from the old man living here, while he by no means seemed inclined to communicate it regarding the surrounding country. He now told us, however, that the branch of the river we last visited came out of a large pond on the east, which had a half-moon shape, stretching away to the south farther than either he or his sons had ever penetrated, and that in a south-west gale there was a heavier sea or swell on this lake than was safe for a small boat. He said also that this pond stretched towards St. George's Bay, at which place I hoped to hear further intelligence of it. On the opposite side of the sound to the old man's house we found an Indian wigwam, with an old Indian woman and her two daughters, one of the latter of whom has a daughter married to one of the old man's sons opposite. The old woman had a kind of moustache tattooed on each cheek, and spoke nothing but Indian, but one of the daughters could speak English. They were busy making baskets and mocassins, and were very neat, tidy, civil people. I bought a pair of very pretty mocassins for half a dollar, made of dressed deer-skin (like Woodstock gloves), and ornamented in front with bits of coloured cloth. The wigwam was composed of a frame of poles like that before mentioned, and covered with large strips of birch bark, kept in their places by other poles resting upon them. The top of the cone was left uncovered to let out the smoke, and the door was closed by a curtain of deer-skin. There was a small fire on the ground in the centre, around which was arranged a layer of small boughs and twigs of fir in a circular fan shape, form-ing a mat to sit down on. They sat some-thing like Turks, with their legs doubled under them. An Indian man, the husband of one of the daughters, was living with them, but was now away in the woods.

August 18th, 19th, and 20th.—Calms and contrary winds detained us in various parts of Humber Sound, and we only reached Lark Harbour on the evening of the 20th. We visited one family about half way down the sound that were much more civil and intelli-gent people than the old fellow's family at the

head of it. Their name was Blanchard, and
the old man had lived here about sixty years,
having settled there before the breaking out
of the American war. He had several sons
that were getting married and beginning to
settle about him. His house, though small,
was neat and comfortable, and he had two or
three small fields under cultivation. They
were just getting in the hay from one small
meadow, and it appeared of good quality.
They had very good currants, raspberries, and
gooseberries in the garden hedges. These
three kinds of fruit we also found at various
places wild in the woods. At one part of
Lark Harbour, where there had been one or
two temporary huts and cleared spots, the
raspberries were in the utmost profusion, and
were equal both in size and flavour to the
best garden raspberries of England. Currants
were found pretty plentifully also, chiefly on
the cliffs, or wherever there was a broken
bank with rocky ledges. They were both red
and black, and of a different species from our
English currant, being covered all over with
small spines like the rough red gooseberry:

the branches, too, had occasionally a soft thorn.
Their flavour was rather harsh, but still very
agreeable, especially when made into puddings.
The gooseberries were more rare, but occasionally we found a small bush, the fruit
being small and very sweet, precisely like the
small rough red gooseberry of England.

August 21st.—Set sail, and tried to beat up
for St. George's Bay against a south-west
breeze, which, when we had made about
fifteen miles, failed us, and for two days it was
stark calm. There we lay rolling idly on the
long smooth swell, with a burning sun and
not a breath of air, and with the land in sight
at a distance of ten miles, through a hazy
atmosphere. On the 24th, finding ourselves
in soundings, having drifted some distance
with the tide, we tied a hook to each end of
the log-line, baited it with a piece of pork,
and in about an hour and a half we had the
deck covered with great cod-fish. There were
fifty-five of them, weighing from five pounds to
thirty pounds apiece, and the men split them
and packed them into the beef-tubs with some
brine. Having once caught one, we were at

no loss for bait, as there was always something in the stomach of one sufficiently attractive for another; or, if his stomach was empty, a piece of that itself generally tempted one of his fellows. The most killing bait was a bivalve shell, with its enclosed animal (some species of *Glycimeris*, I believe *G. siliqua*, which we found in several of them.

CHAPTER IV.

AUGUST 25th.—A light wind during the night
from the west had brought us close up to
Cape St. George, and we made a tack in
order to weather it, when a gale sprang up
from the south-west, and we were obliged to
take in our topsail, and failed in getting sufficiently to windward. We then hoisted the
topsail and made another tack, the sea rising
very rapidly, and barely succeeded in scraping
round the point, which is both shoal and
rocky. There was a short perpendicular wall
of rock at the margin of the water, up which
great seas were now jumping and dashing in

a style that would have rendered our escape
very problematical had we struck on anything, or had any of our gear given way at a
critical moment. Once fairly round, we easily
hauled off into the bay, and then squared the
yards and sailed merrily before it. When we
were about half-way up the bay we saw the
Cleopatra, the frigate on the station, coming
out, at which I was greatly annoyed, as I had
hoped to have met with her, and had now
been twice unsuccessful.   St. George's is a
very fine bay, rapidly narrowing towards the
head, with two straight shores, each of which
affords good anchorage. The only harbour,
however, is just at the head, formed by the
projection of a narrow spit of sand; and even
that seems rapidly filling up with sand, as it
is only near the entrance there is water
enough for vessels, while the rest of the basin
is nearly dry at low water, and is at no place
deep enough for anything but a punt. On
these low sandy shores at the head of the bay
the tide, though not great, becomes very apparent, rising and falling from five to eight feet.
The low spit of sand forming the harbour is
in some places covered with a stunted vegetation of fir-trees. Just at the point, however,

these are cleared away, and there is a collection of wooden houses scattered about as if
they had been taken up from some town by a
tornado, and settled here when it ceased.
The population seemed to be about half French,
and the rest English, Jerseymen, and a few
Indians. There might be perhaps 500 or 600
people at this time, but these are mostly tran-
sitory inhabitants. The French all leave in
November to return in May, and most of the
others retire either to more distant settlements
or to houses in the woods on the opposite
shore during the winter. There were three or
four schooners at anchor, and an old brig,
waiting to take fish to St. Pierre.*

* It may perhaps not be generally known that the fishery
along the whole western coast of Newfoundland, from Cape
Ray round the north point to Cape John, belongs to the
French, or at least that they claim and assert their exclusive
right. The words of the treaty, I believe, admit of some
dispute ; but it is provided that, though the property of the
land is vested in the British crown, neither nation shall
make permanent settlements, and the French shall have the
right of drying fish on any part of the coast they choose.
The provision for non-settlement is practically disregarded
by both parties, as the English settle for their own advantage, and the French connive at or encourage their doing so
on condition that they take care of their stores and fishing
establishments. They also allow the English settlers to
fish within the bays. There is, however, no law nor authority, nor means of establishing any, along this coast, every

August 26th and 27th.—As the Beaufort
had made a deal of water lately, we put her
ashore on the beach here, and at low water
examined her bottom and caulked her a little,
while I strolled about the neighbourhood to
make acquaintance with the people, and get
intelligence of the interior of the country. I
met with an Indian at one of the wigwams,
named Sulleon, a very decent fellow, with a
good character. He told me that he had a
boat on the great pond I had heard of, and
that the nearest point of it was within twenty
miles of the harbour, and that he knew all the
country perfectly well. He said it would take
a week to go and come back again, and he
agreed to go with me as a guide, taking my
four men to carry provisions.

August 28th.—I had got everything ready
for a start this morning, but the weather was
unfavourable, and I felt out of spirits, having

man depending on his own strength to protect himself. A
man of war of both nations goes round once a-year to prevent great disturbances, but, to the honour of the settlers be
it said, there are none to prevent. They all of all nations
seem to live comfortably and peaceably together; and the
only want I heard expressed was a wish for the establishment of schools in St. George's Bay.

got a touch of the influenza, which was very prevalent among the people in the harbour, some of the children having died of it. In the middle of the day there was a thunder-storm, by which the sultry closeness of the air was removed, and the aspect of things greatly renovated. Accordingly, at four p.m. we set out in the boat for the head of the inlet at the extremity of the bay, a distance of ten miles, where we intended to sleep, in order to set off early in the morning. The entrance to this inlet is narrow and shoal, and the tide runs in and out with great rapidity. We with difficulty scraped over the bar, then turned short on the starboard hand where there is a deep channel, the centre of the inlet being one great sandbank. All the head of St. George's Bay seems to be undergoing the process of filling up. The south shores of the bay are composed chiefly of soft red sandstone, the débris of which is washed up the bay by the tide, and sometimes drifted no doubt by the heavy south-west gales. The sand-bank in the inlet extended from its mouth about two-thirds of its length; the upper part had deep water; it was therefore evidently not filled up by the washings of the brooks, but by that of the tide. In the middle of the sand-bank lay an old schooner, which had been wrecked outside and drifted in here. At the head of the inlet we found a brook coming in on its south side, nearly 100 yards across, but with a very shallow bed, and full of rocks and boulders. At the mouth of this was a little strand backed by a slight face of rock or little cliff about ten feet high. Against this face of rock some sticks were laid and covered with bark, the ends being closed by posts, with moss stuffed between them. Some stones reared at one end formed the fireplace, and a small square aperture at the other served for a door. The dimensions of this little cabin were about ten feet in length, six feet in width, and the roof sloped from about five feet high on one side down to the ground on the otner. Here we cooked our supper, and, after drying some fir boughs for a bed, lay down to sleep, there being just room enough for six people lying "heads and points," as the men called it, meaning heads and feet alternately.

August 29th.—We were up with the dawn, and while the breakfast was cooking Sulleon

went to get a shot at some geese, but returned
unsuccessful. We then took the boat to the
north side of the inlet to the mouth of a little
brook, where we hauled her up and secured
her to a tree, hiding the oars and mast in the
woods. It took us then some time to divide
and strap on our several loads. We took in a
small sprit-sail for Sulleon's boat, and a few
tools to make a mast and an oar, a bag of
bread, a small bag of flour, some pieces of
pork, tea, sugar, and molasses, a hatchet or
two, a large boat's kettle and a tea-kettle,
rope, and various other articles. Sulleon had
a bundle of his own of things he was taking in
for his family, as he intended to live at the
pond during the winter, besides his gun and
ammunition. I had a box-sextant, and prismatic compass, a knapsack, containing note-
books, and a lot of biscuit, tea, loaf-sugar, two
or three pounds of raisins, powder, shot, and
balls, a lump of ham, and various other articles,
and strapped on the top of it a blanket and
mackintosh cape and leggings, besides a hammer-bag and hammer, a shot-belt and a double-
barrelled gun, a small telescope, and lots of
things in the pockets of my shooting-coat. We

thus had all pretty good loads. We struck,
near the mouth of the brook, into a little path
leading through the woods, which none but an
Indian would have found, and in about half a
mile came out on a marsh. This had a gentle
upward slope for about a mile, when it inclined the other way down to a skirt of wood
bordering a brook. Scrambling through this,
and walking a little way up the bed of the
brook, we got out on another long flat marsh,
where we called a halt. I here divested myself of my hammer-bag, and also of my
blanket and mackintosh, and distributed them
among the men, as I found my load too heavy.
After toiling for two miles across this weary
marsh we came again to some skirts of wood
and a brook, and made some rapid turns
among underwood and small wet spots to
avoid large masses of wood. We were still
gradually rising, and at length got up a steep
little ascent on to a high level tract of marsh
three or four miles across. The wind was
fresh and cool, moderating a little the great
heat of the sun; and from the place where we
now stood the scenery was very beautiful, but
the extreme toil of the journey took away all

pleasure in looking at it. When we set out
Sulleon declared the distance to be about
fifteen miles, and we intended to sleep that
night on the banks of the pond. It was now
three o'clock in the afternoon, and he told us
we were only half way. It seems ridiculous
to talk of walking seven miles and a half in
about as many hours, but I really do not think
the distance was more, and yet such was the
nature of the country that we all felt knocked up.
We toiled on, however, two or three miles farther, shooting a brace of ptarmigan on our way,
when I found it would be hopeless to accom-
plish the journey before night; and, as Sulleon
said there was a tilt in a wood close by, we
made for it and found a tolerable hut, with the
exception of the roof, which had fallen in.
Relieved from our loads, we soon had the roof
put to rights and slept soundly, with the exception of my luckily waking in the middle
of the night just as the moss of the walls near
the fire was catching from the flames, and thus
saving the house from being burnt down over
our heads.

August 30th.—We awoke rather late this
morning, the sun having just risen. Before

starting we left sufficient provisions for a meal
in the tilt against our return, in case of accidents, as, notwithstanding our heavy loads, we
seemed scarcely to have brought enough to
last us a week. It was a lovely morning, with
little or no wind. A beautiful valley on our
left winded down to St. George's Bay, over
whose expanse we could see blue hills sweeping round in the distance, while on our right
rose wooded eminences with a park-like
scenery on their slopes. But oh how unlike
a park when we came to traverse it! What
would I not have given for a few miles of the
fine turf of old England, or even a heathery
Scotch mountain; anything but the rough,
uneven, scrubby, yet soft and wet spongy mass
of moss we had to stumble through, with
a step between walking and jumping! The
Indian got on most easily, with his toes turned
in, his back bent, and a light yet slouching
kind of gait, dexterously avoiding both the
high knobs and the deep holes, and keeping a
steady pace over all impediments. After
crossing the large marsh we were in, and two
smaller ones, we came to a little circular pond,
round which we went, and, passing through

another small marsh, entered the skirt of woods bordering the grand pond. This wood where we crossed it was about three miles wide. The trees generally were by far the finest * I had seen, and they stood wider apart, and were more open and easy to traverse, so that we could see easily twenty or thirty yards on each side of us. Many, however, had fallen across our path, which was by no means too free from obstruction. About half way through the woods we crossed a brook, flowing east, and falling into the pond. We then ascended for some distance up a gentle slope, and at length from the brow of a small cliff caught our first sight of the Grand Pond. And a beautiful sight it was; a narrow strip of blue water, widening as it proceeded to about two miles, lay between bold rocky precipices covered with wood, and rising almost directly from the water to a height of 500 or 600 feet, having bare tops a little farther back at a still greater elevation. The pond

---

* On the banks of a small brook in this wood I saw a fine *ash*-tree, which to my eyes seemed the same as the common English ash. It was the only one, however, I saw in the country.

stretched directly from us for the first six or
seven miles towards the east-south-east, when
it curved gradually round towards the north,
enclosing the end of a lofty island, and the
water passed out of sight between the hills.
At the foot of the eminence on which we stood
a considerable brook, thirty or forty yards
across, but narrow and full of boulders, ran
down through the woods towards the pond.
Passing down this we came in about a quarter
of a mile to Sulleon's boat, secure and in good
condition, on the bank. We here thankfully
laid down our burthens, and, after resting a
little, I began to feel very unwell, with head-
ache and the feverish symptoms of influenza;
a little tea, however, revived me, and, leaving
the men to get ready the boat and bring it
down the brook, I walked on with Sulleon to
the mouth of the river and the pond. Here
we found some black duck,* and shot several,
which, all hot and perspiring as I was, I fool-

---

* Sulleon saw one of these ducks on the wing a long way
off, and, pulling me down with him into some bushes, he
pinched his nose with his finger and thumb, and then
quacked so naturally that the bird flew right over us, and I
shot him as he passed. Any device which induces wild
animals to come within shot is called " tolling them."

ishly waded after. A little along the south
shore of the lake we found a wigwam, in good
repair, of which we took possession, and began
putting the boat in order, making a mast and
a couple of oars, as there were only two in the
boat. We found, by the wigwam, a birch-
bark canoe, very nicely made, and of an elegant shape, which we launched, and amused
ourselves by paddling about in it. We did
not think it worth while to take it with us,
however, as it required a kind of management
in which none of us but Sulleon was skilled.
We had a grand soup-making this evening of
black ducks, which are excellent eating.

August 31st.—Very unwell this morning,
and the weather very bad; we put off, however,
in heavy rain, and with the wind dead against
us. We were obliged to keep close in-shore,
and the boat, being small and heavily laden,
made but slow progress. Arrived opposite
the end of the island, we found a very good
wigwam, where we determined to shelter, and
where I lay all day with a head-ache, tooth-
ache, back-ache, and all the disagreeables of a
regular cold. At night there was a violent
gale from the north-east, that we expected

every minute either to blow down the wigwam or to bring some of the trees down
upon it.

September 1st.—The lake, at daybreak
this morning, was in a sheet of foam, but the
north-east wind was evidently moderating,
and about seven it shifted to south-west. This
being a fair wind for us, we got our things
into the boat, and set sail. From the point
where we slept the lake divided into two arms,
each from half a mile to a mile wide, enclosing
an island whose summit rose to the same height
as the surrounding hills. Each of the two
arms of the lake, therefore, formed a narrow
ravine, stretching away for several miles, till,
with a gradual curve, they seemed to meet
behind the island.*   We chose the left hand,
or the north-west, of these two arms, as Sulleon
said the other was the best to return by,
having most " harbours."   As we sailed across
towards the island, we were sheltered by the
high land behind us, and had but a gentle
breeze, but when we had entered the arm, and
got into the draught between the precipices of
the mainland on one side and the island on

* This island is, in fact, full twenty miles long.

the other, the wind rapidly increased to a
heavy gale. The greater part of these cliffs,
though very steep, yet had a sufficient slope
for the growth of trees, with an occasional
bare face of rock; the last twenty feet, however, was usually quite bare, plunging like a
sloping wall down into the water, which had an
unknown depth close up to the shore. Sulleon
told me he knew a French fisherman who had
brought three fishing-lines into the lake, and
at a short distance from the end of the island
had failed in finding bottom with them all
tied together; now, as a French fishing-line is
about thirty-three fathoms in length, this
gives a depth of upwards of 100 fathoms.
This would make this narrow ravine as deep
from the surface of the water to the bottom as
it is from the water to the top of the hills. That
the water is of great depth is certain, for as we
drifted before the blast great rolling waves followed us, higher than our heads, and even
calming occasionally the lower part of the sail.
This, although only a small sprit-sail, we were
shortly obliged to reef, leaving a triangle not
much bigger than a pocket-handkerchief; and
even then one man sat with his foot against

the step of the mast, while another supported
it from behind to prevent its being blown out.
Sulleon steered with a paddle, and it was entirely by his skill that we were two or three times
saved from being swamped by the great rollers
that swept after us, and which just curled in
occasionally over the stern of the boat. In
the midst of all this we came suddenly on five
geese, and sent the contents of three barrels
after them as they rose, but the boat rolled so
much that we both missed ; neither, probably,
could we have stopped to get it, had we killed
one. About twelve o'clock we arrived at the
end of the island, which gradually sloped
down into comparatively low ground, and,
putting into a small sandy cove, we went
ashore for a few minutes to stretch our legs
and refresh ourselves. We then again set sail,
and as the lake was now three or four miles
across, and the hills were getting lower, we no
longer had the force of the wind so much concentrated, and sailed rapidly and pleasantly
on till about five o'clock, when we were within
five or six miles of the head of the pond, and
landed in a pleasant little cove on the north-
west side. We must have sailed at least forty

miles to-day, and as we started seven miles
from one end of the pond, and stopped six or
seven from the other end, it gives a length of
about fifty-four miles to the pond, which
agrees with Sulleon's computation of eighteen
leagues, or fifty-four miles. At the point at
which we landed we found the frame of a
large wigwam, which we covered with bark to
a sufficient height to keep off the wind, and,
the night proving fine, we were pretty comfortable. At dusk I walked on along the
sandy beach, but was soon stopped by great
boulders and masses of rock, requiring a good,
light, and steady footing. I sat down on one
of them, and gave myself up to the influence
of the scene. The wind had sunk to a calm,
and the sky was cloudless. Before me lay the
lake, perfectly still, except here and there a
ripple from a stray breath of air creeping
across its surface ; beyond it rose woody hills
getting black with the shades of night; over
these hills and woods there was no track
except the deer-path ; in all the country round
there was no human being except myself and
the few whose voices I could just hear from
the little point, where a small gleam of light

and an occasional spark among the trees be-
trayed our bivouac.   Except this, not a sound
was to be heard,—literally not a sound,—not a
ripple of the water, not a stir among the woods,
not the hum of a single insect, nor the voice of
a single bird.   I believe this utter stillness is
characteristic of all American woods; in Newfoundland it is most remarkable: if you hold
your breath, your ear cannot detect the slightest
interruption to the dead and dreary silence.   It
may, perhaps, savour of affectation, but there
was something most oppressive to my feelings
in this utter absence of sound, and I rose to go
back, when my eye was struck by the most
brilliant aurora I think I ever saw.   A belt
of yellow light rose in the north-east, and
passing just above both the Bears it disappeared in the north-west horizon.   It was not
a perfect arch, but a sinuous band, and it had
a regular onward motion, like that of a
waving ribbon, proceeding from the north-
east to the north-west.   The northern edge, or
base, of this belt was a clear and well-defined
continuous mass of light, while upwards it
faded away into faint parallel rays.   These
rays had no divergence, and seemed to shoot

upwards to a greater or less height from a
certain long narrow base or floor, the plane of
which was parallel to the surface of the earth.
I could have likened it to a long and continued army of celestial spearmen, radiant from
their own light, marching in dense array, with
a regular sweeping course, and gradually unfolding themselves from a distant host massed
together in the north-east, and passing along
in regular procession towards the north-west.
What increased the illusion was, a faint reflection of the central band a little distance on each
side it, but more perceptibly on the outside, or
towards the south, and this reflection followed
the primary band in its long sinuosity, exhibiting the same occasional variations of bright-
ness, and the same upward glancing of the
light.   I am sure the rays proceeded not from
any point in the north, but shot upwards at
right angles to the surface of the earth.   I am
not sure whether I render this description
intelligible, but the effect to me was as if I
was viewing a portion of a sinuous collar of
light, at a great height above, but generally
parallel to some parallel of latitude north of
me, and thus encircling the pole of the earth,

while from this collar perpendicular rays shot upwards. Thus both the arched appearance of the band and the convergence or divergence of the rays, if there were any, would be the effect of perspective merely. At first the greatest mass of light was in the north-east, but it got less as the stream proceeded from it, without perceptibly increasing in the north-west. The effect of this brilliant exhibition in the sky reflected in the still waters of the lake that stretched away beneath it, was majestic in the extreme, and I watched it till its brilliancy began to fade, and at length passed away. There was certainly no noise proceeding from it, or, in the dead stillness of all around, I must have heard it.

September 2nd.—Sailed across the pond to a small brook near its north-eastern extremity, where Sulleon had been told by another Indian that he had seen a considerable bed of coal about three years ago. We traversed the woods along the bed of the brook for about half a mile, when we came on some small cliffs formed by the washing away of the bank. Most of these were covered with gravel and rubbish from above so as to conceal the regular

strata, which appeared to consist of alterna-
tions of yellow sandstone and dark shale or
indurated clay. About a mile up the brook,
however, we came to a small cliff clearly exposed, and there between thin beds of shale
and soft sandstone we found a bed of coal six
inches thick, consisting principally of good
cannel coal. The bed seen by the Indian was
said to be three feet thick, and I have no
doubt was now concealed by the rubbish in
one of the cliffs below. We went some distance farther up the brook, but could find no
more beds, though Sulleon picked up a lump
of good coal six inches thick, and apparently
a part of a larger mass; and as the current of
the brook is very rapid, and its bed rocky, it
must necessarily have come from above.
What I had seen, however, was sufficient to
prove that all these clays and sandstones,
extending through the flat country round the
head of the pond, belonged to a coal formation
containing no doubt good beds of workable
coal. The flat country we had now arrived at
was the same we had reached up the Humber
River, the forking of that river near the rapids
whence we turned back not being at a greater

distance than about eight miles from us. This
flat country was of very considerable extent
towards the north, as from the point where we
slept last night I had noticed two or three low
blue hills rising out of it in the distance,
which Sulleon said were close to the head of
White Bay. The cliffs and hills around the
south-west end of the pond were composed of
gneiss, mica slate, and granite, the two former
of which were the same as the rocks we had
found in the lower part of the Humber River.
For the geological details, however, I must
refer the reader to the report in the Appendix.
The air in the woods was most close and
stifling, and the mosquitoes annoying beyond
all conception, forming by no means the least
of the impediments to steady geologizing in
this country. Arriving again at the beach,
Sulleon and I walked along it, as it was clear
and sandy, and sent the men in the boat to a
point about two miles distant, and in the
centre of the head of the lake. We here
found a fine river coming in from the north,
fifty or sixty yards wide and several feet
deep. The wind had again risen from the
south-west, and the men came in drenched

with spray, having had some difficulty in getting the boat off the shore. We found near the mouth of the river a pretty comfortable wigwam, where we dined, after which, as I had now ascertained the fact of the existence of coal, I wished to return, as our provisions began to run short. The wind, however, was too strong to admit of our making way against it, so we set off up the river in hopes of getting some game. We went up the stream for about three miles, till it became gradually too shoal to admit of our easy progress, without finding anything but a solitary diver, which we shot.

Sulleon told me that the river, which is easily navigable by a canoe, came out of another pond about six miles off. This pond is eight miles long, but near its southern end the river is again found, and, proceeding up it, three more ponds are met with, each about six miles long, and the last of which is about sixteen miles from Hall's Bay. The Indians, carrying their canoes overland from this pond, come in about half a mile to another brook, down which they proceed to Hall's Bay. Returning to the grand pond, about three miles west of the mouth of the river, an equally large river, if it be not considered the same, runs out of the pond into the Humber River, being the branch of that river from which we last turned back in our former excursion. The distance along the winding of this branch he gave as eight or ten miles, but he said it was too full of rapids to admit of the passage of anything but a canoe, which could be lifted out occasionally and carried.

On going back to the wigwam we shot a couple of " twillecks," a grey long-legged bird, about the size and shape of a plover, that frequents the shores of the lakes and arms of the sea. These two, with the small duck or diver, served us as dinner and supper that night, and breakfast the next morning, with the addition of tea and a cake of bread each. This was not a very plentiful repast, but with their usual improvidence the men had wasted the provisions while they lasted, and now were likely to fare but poorly unless the wind moderated.

September 3rd.—At three o'clock this morning, the wind having moderated, and the moon being high and bright, we got our things into the boat and rowed off. It was dreadfully

cold till after the sun rose, and then the wind
freshened. We held on, however, hugging
the western shore of the lake, and by eight
o'clock we calculated we had made fifteen
miles, when we landed to breakfast. The
cliffs here consisted of red marls and sandstones, but upon nearing the island we came
upon slaty rocks again. As soon as we opened
the channel on the north-west side of the
island, the wind met us with increased violence, so, hoisting the sail of the boat, we crossed
over to the entrance of the opposite channel,
where we were more sheltered. We here
landed to examine some small ponds in the
wood in hopes of finding geese, but were dis-
appointed. We got, however, a quantity of
berries, especially a kind the men called
" squash-berries," a bright red berry, the size
of a currant, growing on a straggling bush six
or eight feet high. They were pulpy, sharp,
and juicy, and not unpleasant. I had brought
a pointer with me called Bell, and as we had
lately lived chiefly on water-birds, whose
bones she refused, I had been obliged to share
my biscuit with her. I taught her, however,
now to eat berries. Soon after entering the
channel between the island and the main we

came to a narrow part where two sandy points stretched opposite to each other. This, Sulleon said, was a favourite place for deer to swim across, and as we found a good wigwam we camped here for the chance. We found on the sand abundance of the fresh tracks of these deer or cariboo, one of which was actually not dry, the animal having, apparently, come out of the water and gone off at our approach. We waited in vain, however, for more, but luckily shot a couple of divers, which served us for supper. The men's bread was now all gone, and we had nothing but tea and our guns for support. I had, however, kept in my knapsack a private store of half a dozen biscuits and a piece of ham, against an emergency, as also a bunch or two of raisins, and we had still a little flour left in the bag, for thickening our soup.

September 4th.—We put off at daylight, and rowed some time against a head wind, the cliffs on each side getting gradually steeper and more lofty. At eight o'clock we found we had barely made three or four miles, the point we had left being still visible. We were now on the island side of the channel, and,

coming to a little strand where there was a
small wigwam, we landed to breakfast. It
would evidently never do to go on at that rate,
and, if the wind continued or increased, we
might be three or four days in reaching the
south-west extremity of the pond, where we
had left some provisions. Sulleon said this
island was a good place for deer; accordingly
I produced the remaining stock of my provisions, and, sharing them equally among
all hands, we determined, after having a
tolerable breakfast, to hunt for a cariboo.
Leaving two men with the boat, the two
others, Sulleon, and myself, set off with our
guns. After a stiff and toilsome climb up the
woody precipice we arrived at some open
marshes at the top, sloping gently towards the
centre of the island, where were several ponds,
one about three miles long. There were
plenty of deer-tracks, but no deer visible.
Among some craggy rocks, however, we found
an abundance of bilberries, with which, after
walking some hours, we partly satisfied our
appetites, and Bell learned soon to pick them
for herself. We then separated; Sulleon went
down to the pond, while I went up to some

rocky eminences to try to find some ptarmigan, the utter absence of which rather surprised us. After hunting some time in vain, I heard a shot, and shortly after another. Running was out of the question, but I made what haste I could across the marsh to the knoll where I had left the two men stationed with my rifle. They had heard no shot, but on looking about we saw Sulleon sitting on a rock about half a mile off, and on our coming up to him he said, with all the unconcern in the world, " I hab killed a fat buck." Following him down through a thicket into a little marsh, we found the fellow lying on his back, a great beast as big as a small heifer, and shot right through the heart. Sulleon had seen him swim across the pond in the centre of the island, watched where he landed, and stole down through the tangled woods, as none but an Indian can steal, till he came within thirty yards of the beast without disturbing him. We shortly had him skinned, embowelled, cleaned, and cut up. Sulleon being used to it doubled up one side for his load, which could not be less than a hundred-weight, my two men took a quarter each, while I made a bundle of the lower jaw with the tongue, the heart, the liver, and a good portion of the tripe, &c., selected by Sulleon, and used by him for different purposes, all wrapped up in the skin. We were now three miles from the boat, the island being five miles across, and had a pretty stiff march across the marshes and barrens under a hot evening sun and our sufficiently heavy loads. We were, however, all in high spirits; and on reaching the precipitous wooded bank of the lake we started on a kind of race to see which should be first to carry the news to the boat. We consequently all lost our way in the woods, all had tumbles, and my bundle fell from my back, and its contents got loose, by which means I arrived last at the camp. Great was the frying, roasting, boiling, frizzling, and stuffing of maws, Bell being by no means the least happy of the party, as venison liver and bones met with her decided approval. Great also was the talking and laughing; and when Sulleon told us in his quiet grave manner that, in coming down the woods, "I fell, my God, at least ten yards!"

one would have thought it an excellent sally
of wit by the peals of laughter with which it
was echoed,—

"For mountain lads can lof at turning of a straw."

Our only drawback was the want of bread and
salt, but nothing in this life can be perfect.
There was a fine aurora at night, but the rocks
and woods above and opposite to us inter-
cepted our view of it. Soon after there were
some heavy showers of rain, but

"After that fitful supper we slept well."

September 5th.—At dawn this morning we
stowed our meat and baggage in the boat, covering it with the deer skin. The morning was perfectly calm, but foggy, with occasional showers.
I noticed before we went off that a piece of
wood which we threw into the water floated
regularly alongshore in the direction we were
going, or towards the south-west end of the
pond. Sulleon said there was often a tide in the
pond after a high wind. This is no doubt caused
by the banking up of the water at one end
from the pressure of the wind. There is only
one brook runs out of the pond,—namely, that

at the north-west corner, which goes to join
the Humber, whose stream is quite inconsider-
able compared with the body of water in the
lake, and therefore could produce no current
either way. Sulleon told me this morning
that he had spent two or three severe winters
on the banks of the Grand Pond; and that
while about fifteen miles of the north-east end
of it was frozen over completely every winter,
he had never seen the southern part near the
island "set fast," or sufficiently frozen over to
enable him to cross. This fact goes in corrobo-
ration of the statement of the enormous depth
of the south-west end of the pond, where the
water is narrow and the banks lofty, compared
with the north-east end, where the lake is
broader and the shores flat. The bottom of the
lake at its south-west end must be 300 or 400
feet at least below the level of the sea. One
is puzzled to understand the formation of such
a deep narrow ravine among these hills, the
height of the hills above the sea being by no
means great. But, indeed, these narrow crevices are frequent in Newfoundland, and more
frequent beneath the level of the sea than
above it, as the number of narrow inlets eating

into the coast will testify, they having often a
length of ten or twenty miles, a breadth of one
or two, and a depth of upwards of 100 fathoms,
while the cliffs on each side of them are often
not more than 100 feet in height. Two faults
near together throwing down the intermediate
piece might explain these clefts, but their
course is often winding, while we should conceive that two parallel faults of that magnitude
would almost necessarily be straight.

The morning continuing quite calm we
rowed steadily on, and reached the south-west
extremity of the lake between two and three
o'clock. Sulleon then proceeded to cut some
of the meat of the venison into long strips,
which he hung up in the wigwam on sticks
over the fire, and meant to leave it there till
we came in with his family. The Indians dry
the venison in the smoke in this way till it
assumes the appearance of old leather, and
then use it without cooking, carrying it with
them in their hunting excursions. Sulleon
told me that he himself had killed upwards of
fifty deer during the last winter chiefly by
watching them swim across the lake at par-
ticular points, and then chasing them in his

canoe and stabbing them in the water with a
short iron spear, which is always kept in the
canoe for that purpose. The English settlers
have a more cruel method, as when they catch
a deer in the water they break his back with a
hatchet. The animal can then still swim,
and they drive it before them; but the moment
it gets on shore it falls, when they dispatch it.
This is simply to save themselves the trouble
of hauling it into the boat or towing it after
them, and I was glad to see Sulleon join me
in expressing abhorrence of such unnecessary
cruelty. When Sulleon had taken what meat
he wanted we took the bones out of the remainder and packed it up to carry out. We
then amused ourselves in firing at a mark
and paddling the canoe. This canoe, made
merely of birch bark sewed neatly over a light
frame of thin wood, looks the frailest of all nut-
shells. It will live, however, they say, in a
heavy sea merely from its lightness and buoy-
ancy. It was about ten feet long and two feet
broad, sharp at both ends and round-bottomed,
and I have certainly no wish ever to experience
its qualities in a heavy sea or a gale of wind.
I was told that Indians had crossed in them

from Cape North in the island of Cape Breton to Cape Ray in Newfoundland, a distance of sixty or seventy miles, part of it out of sight of land; but I should greatly doubt the truth of the statement.

September 6th.—In setting out this morning we were delayed a good deal by getting the boat aground at the mouth of the brook, and having to drag it up into a place of conceal-ment. It was thus half-past seven o'clock before we had fairly got our loads on our shoulders and set off into the woods. The morning was heavy and dull, with a driving fog and an easterly wind, and just as we set off it began to rain. The first plunge into the bushes wetted us to the skin, and was like step-ping into a cold bath. Every stick was so wet and slippery that it took us nearly four hours to get through the three miles of wood, being delayed, however, twenty minutes by one of the men dropping his bundle and losing his way. On getting out into the barrens and marshes we found the wind very cold and driving a deluge of rain upon our backs. With great care we managed to keep our guns dry for some time; and coming down on a

little pond we shot a black duck. Presently
after we came on a covey of ptarmigan, quite
tame, and apparently unwilling to get up; of
these we shot three brace. At one o'clock we
reached the tilt in the woods where we had
slept as we came in. We were now thoroughly drenched. Determined to make a
bold push to get back to St. George's Harbour
and our little vessel to-night, I only allowed
one hour to cook and eat dinner and tea and
start again. Accordingly, at two o'clock we
were again on our way, and toiled vigorously
over the soaked marshes, and through the cold
wet woods. We saw abundance of ptarmigan
and several flocks of wild geese, but by this
time our guns were useless from the wet, and
we had no time to spare had they been dry.
Between four and five we made a simultaneous
halt beside a small brook, and one and all
declared we could go no farther without something to eat. We accordingly unpacked some
cold venison; and I got out a lump of sugar I
had left, and dissolving it in water made one
of the most pleasant and refreshing draughts
that can be taken on a hard march. Sulleon
said venison was the weakest of all meat,

meaning that one became sooner hungry again
after eating it. Salt pork, he said, satisfied
the appetite longer than anything else. In
passing over the last two or three miles of
march the men began to flag very much, as
our bundles and clothes being saturated with
wet increased our burdens somewhat. Sulleon
also complained of a pain in his stomach from
over-fatigue. We kept steadily on, however,
and just as it was beginning to get dusk
reached the last ridge and the skirt of the last
half mile of wood. We had some difficulty in
hitting on the little track among the bushes at
the skirt of the wood, and had we missed it all
our labour would have been in vain, and we
must have passed the night where we were.
Luckily, however, Sulleon smelt it out, and
traversing the woods as rapidly as we could
we reached our boat just as it got too dark to
proceed any farther. Here we gladly threw
down our bundles, hauled the boat into deep
water, stepped the mast and bent the sail, and
were soon rapidly drifting down the inlet
before a strong east wind. For the first quarter of an hour we were in high spirits, but as
the cold wind blew through our wet clothes

the talking and laughing gradually died away, and we sat shivering in the boat in silence. Though wrapped in my blanket and cape, I think I never passed a more wretched time of two hours and a half than this, as, crouched in the stern of the boat, I in vain peered through the darkness to get a glimpse of some light on shore or the form of some land by which to mark our progress. At length we passed into the harbour, struck the side of our vessel, up which I could hardly climb for numbness, and astonished Gaden the skipper with our appearance at this time of night. A glass of brandy all round, however, prepared us for a supper of hot venison; and taking off my clothes for the first time for ten days, I soon slept off the fatigue between the blankets in my berth.

September 7th.—Very busy all day writing out notes, drawing maps, &c. Just by the south-west extremity of the Grand Pond is a large bare round-topped hill, plainly visible from St. George's Harbour, from which it bears due north-east, true bearings. This hill Sulleon called Hare Hill, from the abundance of hares upon it. It must be about 1000 feet in height. None of the land we passed over

rose more than 400 or 500 feet above the sea,
and a road might readily be constructed across
it. If necessary at any time, by means of a
few canoes, an easy communication might then
be kept up by means of the ponds and rivers
between St. George's Bay and Hall's Bay.
A hill on the south-east side of the pond,
about ten miles from its north-eastern extremity, called by Sulleon Ories Hill, is a conspicuous object. The contrary winds and our
shortness of provisions would not allow me to
visit it, which I regretted greatly, as from its
summit Sulleon said the Red Indian Pond
might be seen a day's journey to the east, and
a very lofty hill forty miles off (by his esti-
mate) in the south-east, which may probably
be the Mount Misery or Jamieson's Mountain
of Mr. Cormack. Were the western side of
the island settled, the banks of the Humber
and the north end of the Grand Pond would
be by far the most favourable spots for an
inland population. The soil is richer, and the
inland communication might be greatly extended by means of a few roads between the
ponds and rivers.

CHAPTER V.

Raised Beach—Crab's River—A bed of Coal—South side of St. George's Bay—A Storm—Gypsum beds—Red Indians—Adventures with—Their Superstitions—Coal—Wreck of the Onondago—Narrow Channel.

SEPTEMBER 8th.—Sailed from St. George's Harbour, and examined Indian Head and the north shore of the bay. It was a most lovely day, and the chain of hills that sweep from the south-west round the head of the harbour, and pass off towards the north, showed to great advantage, with their patches of light and shade, their sunny peaks and dark hollows.

September 9th.—Landed in Ship Cove. The whole of the north shore consists of low perpendicular cliffs of yellow magnesian limestone, very straight, and with scarcely any indentation. Just east of Ship Cove is another small cove at the mouth of a little valley,

down which a small brook runs out to the sea.
The valley is full of clay, mingled with boulders, on the top of the crumbling cliffs of which
I found a number of loose, partly decomposed
shells. They consisted of *Mya arenaria*, a
species of *Pholas*, and one or two small species
of *Tellina*. The same species of shells are
abundant in different parts of the bay, but I
did not see any in this cove, the beach of
which is covered with large pebbles. For
want of a spade I could not ascertain whether
the shells were buried in the clay to any
depth, and what I found on the slope of the
bank might have been brought by birds, and
buried slightly by clay washed down by the
rains.   It was, however, the only spot in
which any kind of evidence could be found to
show a recent elevation of the coast of any
part of Newfoundland which I visited, and
here the evidence was uncertain, to say the
least of it.   The shells generally, on the coast
of Newfoundland, are rare, both in species and
individuals, and are only found in such places
as St. George's Bay, and the few inlets in
which there is shoal water, and a sandy or
muddy bottom.

In the evening we beat up to the southern shore of the bay, and anchored under a headland, and during the night it blew a very heavy gale of wind from the south.

September 10th.—Went to the mouth of Crab's River, in order to explore for the coal reported to exist on the south side of St. George's Bay. This is a considerable brook, but has only depth of water enough to admit even such a small craft as ours at spring tides. There were two houses near it, in which were English settlers; and we got two men to aid us in warping the Beaufort over the bar into the little cove at the mouth of the river, where, when she was secured, there was but just room to row a boat round her. The cliffs here consisted of soft red and white sandstones, and red and variegated marls. The top of the cliffs, for about a hundred yards in width and a mile or two in length, was occupied by a fine short turf, like that of a sheep-down. At the back of this stood the usual dense wood, but the relative number of the different species of trees was not the same as on the slate rocks. There were more aspens, and other branching trees, and fewer firs and

larches. The country, in the interior, was
gently undulating for some distance, and more
like some parts of England than any I had
yet seen in Newfoundland. The people, too,
seemed more agricultural than in other parts,
with poultry and cattle about them. They
informed me that the only place in which coal
was to be seen was in the banks of a brook
about three miles above them, and some miles
from the coast.

September 11th.—Soon after daylight we
set off in the boat for this other brook, and
landed at its mouth. It was a very pretty
little spot, with green meadows on each side
of it, and two or three neat clean houses clus-
tered under the shelter of a rising bank
covered with green turf. Geese were feeding
on the grass, ducks and poultry were scattered
about, and a few cows and some sheep gave it
all the appearance of a pastoral scene at home.
There was actually a fence and a stile to get
over into a small field with a footpath across
it, a sight quite refreshing to the eyes after so
many months looking at wild rocks and
dreary woods. Mr. Morris, the patriarch of
the settlement, came and invited me to sit

down to breakfast with them, when I found
plenty of fresh milk, eggs and butter, hot
bread, excellent tea, and a snow-white table-
cloth.   It really seemed a little paradise.
Stephen Shears, his son-in-law, who had a
house close by, volunteered to "pilot" us up
the brook to a place where he had himself
seen coal.   From the rising ground behind
the houses the view was very beautiful.   A
tract of low undulating land, covered with a
rich sea of wood, stretched away into the interior for fifteen or twenty miles, and was
backed by a range of blue hills in the horizon
that rose towards the south-west, into a still
loftier range running to Cape Ray, while
towards the north-east they gradually died
away, and coalesced with the hills at the head
of the bay.   The wood was not of the sombre
hue so generally seen in the country, but was
patched with the light green of the birch, and
what they here called the "wych hazel," the
"barm," and the "aps."   I think, also, there
were ash-trees, but could not be sure.   The
little rich-looking valley of the brook, with its
bright waters winding away into the woods,
completed a most lovely, and almost English,

picture. We passed over the hill, and descended into the valley of another small brook close by, to the west of the former one. Up this brook we proceeded. Its width varied from 20 to 100 yards, and it was generally shallow, and its bed full of enormous boulders, much rounded. These boulders consisted chiefly of gneiss and granite, and Shears said they were moved and brought down by the ice in the spring. We had frequently to cross the brook to take advantage of the various little banks of shingle or sand. The rest of the journey consisted of a compound of wading and jumping from one slippery block to another. We had, however, no burdens, except a pick-axe and shovel, and a bag to bring away anything we might find. The banks of the brook, at first, consisted of red and white sandstone, and marls dipping north-west, or towards the sea. These marls and sandstones were exactly like the new red sandstone of England; and as they inclined from the country where the coals were said to be, I was marvelling at the exact coincidence of the relations between the two formations in such distant countries, and inclined at once to set down these red sandstones as new red sandstone. After proceeding about three miles, however, I came upon some perpendicular beds, and after that found all the red sandstones and marls dipping the contrary way, or towards the coal country. After proceeding some distance, these red marls and sandstones began to alternate with dark blue clays containing gypsum, brown sandstones, and conglomerates, and presently with dark grey indurated shale, or clunch. The grey and yellow sandstones and shales then began to preponderate over the red ones, and occasionally a bed of black, bituminous shale was observable. I then began to perceive that the red sandstones and marls composed, in fact, the base, or lower beds of the coal formation ; and, to make a long story short, we came, at a distance of eight or ten miles from the shore, on a bed of coal at the top of a small bank, resting on grey clunch and sandstone, from under which, at no great distance, a bed of red marl cropped out. We immediately set to work with pickaxe and shovel, and after filling our bag with the best pieces of coal, we made a fire on the beach, and had a famous

blaze with coals of our own digging. The
coal is of a good quality, and a good deal of it
is cannel coal. The bed exposed was between
two and three feet thick, but as the top was
wanting, it may belong to a still thicker bed.
The bank immediately beyond it ended, and
thus cut off its further extension, neither was
there any appearance of it farther up. Shears
said this was the only spot he knew in which
coal appeared at the surface of the ground.
There is, of course, no doubt of there being
more beds in the vicinity, and of the probability of all the centre of this low district
being occupied by a productive coal-field. We
certainly cannot tell what the future may
bring forth, but at present this coal-field must
lie entirely dormant, as, from its distance from
the coast, it cannot compete with the extensive coal-fields of Cape Breton and Nova
Scotia, now in work. After eating a lunch
we set out on our return, and got back just at
sunset wet up to the middle, although the day
had been a beautiful one. They gave us some
tea, and soon after dark we got into our boat,
pulled through a rolling surf, and reached
Crab's River in about an hour.

September 12th.—Stephen Shears came down this morning, to beg me to write some letters for them, to their friends at home. He had been an apprentice in Devonshire twenty-five years ago, and when about sixteen ran away and came out here. He served his present father-in-law as a hired labourer some time, till he gradually accumulated a little money. He had now eleven cows, three oxen, and a bull, twelve sheep, a house at the brook for a summer residence, another house in the woods for the winter, where he had also a garden : he had also a few roods of ground cleared, which he had sown this summer with French wheat, which, he said, thrived very well. A few years before he had married his old master's daughter. His father-in-law, Morris, himself originally came out as he did, without a farthing, and now they were all happy, comfortable, and independent. The only want which Shears expressed was that of books and schooling, for himself and his children, and he begged me very earnestly, if ever I came that way again, to bring him some. I had none with me, but books of reference on scientific subjects, or I should certainly have

given him some. He says the snow here
generally sets in about three weeks before
Christmas, and breaks up in the beginning ot
April. The climate, during the summer, is
very fine, and certainly, while I was in St
George's Bay, nothing could exceed the beauty
of the weather.

Existing treaties will, I suppose, preclude
the regular settlement of this neighbourhood,
but as far as its natural capabilities and resources go, St. George's Bay and its neighbourhood is by far the most inviting part of
Newfoundland. It is, indeed, the only part in
which agriculture could flourish so as to become part of the resources of the country, and
is likewise the only part which has any mineral
wealth to boast of. Had the western shore of
the island been the eastern, it would, before
this time, have contained a populous and flou-
rishing community.

At twelve o'clock to-day, being the top of
high water, we managed to scrape out over
the bar, getting one knock, however, from
a hard rock, some great boulder, probably,
brought down by the ice out of the river.
We then stood out into the bay, for Cape

Anguille, but, there being a light wind from the west, did not make much way.

September 13th.—Light breeze from east-south-east, with clouds and heavy rain. Doubled Cape Anguille, and entered by a rocky channel into the little harbour about three miles south of it, formed by a small island called Codroy Island. There were two small schooners at anchor there, and a few huts on the island belonging to French fishermen, while on the shore were a few English settlers. It rained hard all day, and the wind gradually increased from the east. After dark it was blowing a gale from the south-east, which seemed to increase every minute. We got out another anchor, and about seven the two schooners dragged their anchors, and one of them drifted down upon us, but luckily brought up before she fouled us. As Gaden and I were talking in the cabin, we also began to drift, one of our anchors having loosed its hold. On coming on deck we found it pitch dark, and the wind blowing so furiously that we could hardly stand. I could just make out the outline of the land close by us, and saw, by its apparent motion, that we were

drifting rapidly. We were obliged to stoop
below the bulwarks, and shout in each other's
ears to make ourselves heard. There was
no sea, as the wind blew off-shore, but the
whole surface of the water was a sheet of
white foam, which the blast tore up and
shot into our faces, making them smart as
if the spray were crystals of salt. We veered
away all our cable, and when within fifty
yards of the point of the island forming the
harbour, the water getting shoal and the
bottom more sandy, our anchors again bit
well, and we held on. Here we determined
to run her ashore, cut away our masts, or
even take to the boats and abandon her, rather
than be blown out to sea in such a night; and
I went to pack up my note-books and other
light valuables in case of necessity. We held
on here for some time, but swung occasionally
in-shore, and presently began to strike heavily.
By shifting the yards and the helm we sheered
off into deeper water, but still struck once or
twice. I turned into my berth at ten o'clock,
but left my pea-jacket, with my watch and
note-books, ready for a start should it be
necessary. Between twelve and one the wind

moderated a little, and at one o'clock, just at high water, the wind suddenly shifted into the north-west. They took this opportunity of getting the vessel back into a safe berth, which they effected with some difficulty. For the remainder of the night the wind blew from the north-west as hard as it had done from the south-east, and we had every reason to congratulate ourselves on being where we were.

September 14th.—The wind a little moderated, but still blowing hard from the north-west. The channel by which we came in was now a wide-spread bed of breakers. Standing on the corner of the little island and looking towards Cape Anguille the sight was magnificent. Long waving ridges of sweltering water, with flickering tops like small mountain peaks, came rolling successively towards the land, rose and curled over when they touched the rocks, and fell in regular cataracts of foam like the falls of some great river. I at one time could count from six to seven of these great ridges of water, for they could hardly be called waves, each two or three miles long, coming in regularly one behind the other. It was not possible to land on the main, as the sea curled round each end of the island into the little harbour and caused a heavy surf on the opposite shore. There were several small fishing-boats in the harbour sunk at their anchors.

September 15th.—Fine, with light air from the south-east, so could not proceed to the river which I wished to visit. Found the stock of one of our anchors broken. Landed on the main and examined the cliffs, in which there was a great quantity of excellent gypsum in beds and veins. Heard of a large vessel being ashore near Cape Ray, sails all set, and no one in her. Many boats lost along the coast.

September 16th, 17th.—Wind south-east, so could not get out of the harbour. A schooner from the Magdalen Islands came in that had been out in the gale. They said it was the heaviest they ever knew. Her deck-boat was smashed from end to end by a heavy sea.

September 18th.—Calm. Towed the Beaufort out of the harbour into the mouth of the river, called the Great Codwy River. A large harbour is formed by a long spit of sand across the mouth of the river. At low water, however, most of it is dry, with the exception of

the channel of the river, which has a depth of
ten or twelve feet. Two or three English
families, with cattle, reside on the south side,
and one or two Irish, with some pigs, on the
north. On the sandy point were three Indian
wigwams. All the men had left the place and
gone down to the wreck, except two old Indians
and a young one, who was a cripple. This
young Indian was about thirty, a fine intelli-
gent fellow, with a noble countenance and
piercing black eye. About two years ago he
had fallen, while carrying a barrel of flour,
and his back was so severely injured that he
has not been able since to walk without
crutches. His wife was an exceedingly pretty
woman, with a Grecian countenance, dark but
ruddy complexion, and a sweet smile. She
spoke no English, and was very modest and
reserved. They had been married ten years
and had three children. Her husband gave
me much information as to the interior of
the country, and an animated description of
his having once seen a Red Indian fishing in
a river in the interior. This was about twelve
years before, and since that time he had
neither seen nor heard anything of the exist-

ence of the Red Indians, although before his
accident he had traversed every part of the
western side of the island.

When Newfoundland was first visited by
Europeans it was peopled by a race of Indians
who, from the colour of their skins and from
the red ochre with which they smeared themselves and everything belonging to them, were
called Red Indians. They were no doubt the
same aboriginal race as the red men of the
continent of North America. They were at
first friendly, but in process of time a deadly
enmity arose between them and the whites, so
that neither spared the other. There is too
much reason to believe that the greatest atro-
cities were committed upon these Indians by
the rude fishermen who settled in different
parts of Newfoundland. They complained that
the Indians stole their boats and nets, and they
shot them down whenever they saw them.
Stories are current too of persons having gone
into the interior on purpose to hunt the Red
Indians, and shoot them for the sake of the
furs and skins they had about them. Thus,
partly from actual slaughter, and partly from
the best parts of the coast being occupied by

Europeans, the Red Indians seem to have decreased rapidly in numbers. At the beginning of the last century a body of Mic Mac Indians, partly civilized and converted to the Roman Catholic faith, either came or were sent from Nova Scotia, and settled in the western part of Newfoundland. These were armed with guns and hunted the country, making great havoc amongst the game. A quarrel soon arose—perhaps on this account—between them and the Red Indians; and Sulleon gave me a confused account of a battle that took place between them at the north end of the Grand Pond about seventy years ago. In this the Red Indians were defeated, as they were armed only with bows and arrows, and, according to Sulleon's statement, every man, woman, and child was put to death. About fifteen years ago attempts were made by a society, formed in St. John's, to open a communication with the remainder of these Red Indians, but without success; and Lieutenant Buchan, with a party from a man of war, having surprised and surrounded a tribe of Red Indians, near Exploits Bay, tried to conciliate them, and invited two of them on board his vessel, leaving two of his marines as hostages. The Indians, however, escaped, and returning to their encampment, he found it deserted and the bodies of his two marines lying on the ground without their heads. Subsequently to that, one or two women were captured, and one lived in St. John's some years. No Red Indians, however, had been seen in any part of the island for the last ten years, and I believe the last remnant of them have either perished or have passed over to the Labradore. Such, however, is not the opinion of the settlers in the distant outports. While on the Grand Pond my men were in continual dread of meeting the Red Indians, and their feeling towards them may be estimated by their open avowal that if they saw one they should fire at him as if he were a wolf. Their astonishment was really amusing when I told them that, if they did meet one and shot him without provocation, I should do my best to have them hanged on their return to St. John's. They evidently looked upon them as wild animals of a pernicious character.

The Mic Mac Indians inhabiting Newfound-

land reside chiefly on its western side, wan-
dering from Fortune Bay to St. George's, and
thence to White Bay and the Bay of Exploits,
living in the winter on the products of the chase,
and in the summer joining in the fishery, or
getting other desultory employment. Their
number probably does not exceed a hundred
families, and they bear on the whole a good
character. A few of them are worthless vaga-
bonds, and, especially when intoxicated, are
not to be depended on ; but the greater number are honest, well-meaning fellows, and
when well treated are faithful and steadfast.
Sulleon and the cripple at Codwy River were
men of a superior stamp, both morally and in-
tellectually ; and had the latter been well and
unhurt, I would rather have had them for
companions in the interior than any of the
Europeans I met on the coast. They are all
Roman Catholics ; and Sulleon's daughter having been recently married to a young Indian
when I was in St. George's, they had gone
across the country to White Bay, a short time
before our expedition, to have the ceremony
performed by a Roman Catholic priest, who,

they heard, was travelling upon the north coast of the island.* I suspect, however, that, like most other people, they still preserve some of their old superstitions. Sulleon, for instance, would not allow one of my men to fire at a jay with his gun, as he said the gun would be spoilt, and get gradually worse after killing the jay, until it burst or would no longer shoot straight. He would not give his reason for this notion ; and I could see that upon this and several other points he held notions which he did not like to express. From all I could hear they are a very moral people, being especially strict with regard to their women and marrying at a very early age.

September 19th to 24th.—Detained in Codwy River by contrary winds. Some of the men coming back from the wreck told us she was the Onondago of Cork, loaded with timber. The day after the gale she was seen in communication with a brig, which took out her

* All the inhabitants of the coast north of Cape Ray, both Indians and Europeans, marry by drawing up papers before witnesses, binding upon each party to have the ceremony performed on the first opportunity that presents itself. As there are neither clergy nor magistrates, this is, of course, their only plan, and is tantamount to the Scotch law.

provisions and crew and let her go, after which she drifted in-shore. I heard reports of coal having been seen by the Indians up Codwy River, and tried to get more exact information and persuade one of them to go as a guide. Several professed to know where it was, and offered to accompany me, but when the time came always drew back. One Indian woman, middle-aged, and apparently intelligent, began talking about the " Indian king." She said that they had a " king," who resided in Nova Scotia, that the whole country rightly belonged to him, and that he must be consulted before any of them dare give me any information. Another said he would go and show me if I would give him a hundred pounds, as I might get thousands or millions. I was told also that an Englishman, named Gale, who lived on the south side of the river, and who had the reputation of being very rich, had strictly charged the Indians to give no information. Those Indians who resided all the year round were sometimes dependent on the English settlers for food, and dare not disobey them and the latter were particularly churlish and uncommunicative, especially old Gale. They

feared, it seems, that mines would be established, and that thus the neighbourhood would be
regularly settled, and not only their own trade
and authority interfered with, but taxes and
customs imposed. All along this shore I was
looked upon with great suspicion by many of
the European residents as a Government agent,
and it was believed that my ultimate object
was the enforcing taxes and custom-house
duties. After being amused for two or three
days by vague promises on the part of the Indians to come and show me where the coal was,
I was privately assured by Peter, the cripple,
that none of them had ever seen the coal themselves, and only knew of it from the report of
an Indian, named Morris Lewis, who described
it as in a lateral brook of this river, at a distance of about thirty miles. I explored the
river for about ten miles, and found a lovely
valley between some gently rising ground on
the one hand, and the bold and precipitous
range of hills running down to Cape Ray on
the other. The rocks were the same as those
I had seen near Crab's River, and dipped towards the interior of the country, in which
direction there is, therefore, no doubt of the

existence of coal.   There were several strips of grass along the banks of the river, which had been cut and made into hay, and on some small islands were a few very good potato-gardens, belonging to the Indians down at the point.   The day before we left I bought a good dog from one of the Indians for ten shillings. Newfoundland is one of the worst places in the world for getting a good, or at least a good-looking, Newfoundland dog.   In St. John's and its neighbourhood they are the most ill-looking set of mongrels that can be conceived. In the more distant ports, however, the breed has been better preserved.

September 24th.—At length we got a slight breath of easterly wind, just sufficient to send us out of the mouth of the river, when a north-north-west breeze came up, which was also the best wind for sending us round the Cape. Cape Ray is a very conspicuous object at a distance, in consequence of the bold flat-topped ridge of hills that comes down to it, ending in two or three detached conical eleva tions.   From the foot of these about a mile of flat ground stretches out to sea, forming the point of the Cape.   In a small cove on the

north side of this point, high on the rocks, lay
the wreck of the Onondago. She appeared to
have been a fine ship of 700 or 800 tons ; her
hull seemed not to have sustained much in-
jury, but her masts were now gone, and she
was of course stove in below by the sharp rocks
on which she lay. Two small sloops, two or
three fishing-boats, and several punts lay alongside, and their crews were on shore getting out
her cargo, piling the deals and other timber
of which it consisted, and loading their own
vessels with all imaginable coolness and deli-
beration. She was evidently considered fair
game, and 'first come first served' seemed the
order of the day. The dirty little vessels
around her put me in mind of a parcel of
wolves preying on the carcase of a race-horse.

On passing the Cape there was a very striking change in the aspect and character of the
country, but the change was by no means an
improvement. The range of hills before mentioned, whose steep north-western face over-
looked the rich and apparently fertile country
of St. George's Bay, sloped more gradually towards the south-east, into a wild, bare, rugged
mass of broken rocks inclining gradually to the

sea, clothed only with moss and small berry-bearing bushes, and furrowed in every direction by narrow, and abrupt ravines. This rugged and broken country was evidently continued with the same character beneath the sea. The shore is indented with creeks and inlets in every direction, and a fringe of rocks and rocky islets, two or three miles broad, runs along it. Keeping alongshore we got entangled among sunken rocks before we were aware of it, as the water was quite smooth. A boat just ahead of us, however, left Codwy River a little before us, bound for Port aux Basques; and by keeping a diligent eye upon her motions, and following her winding course, we managed to keep the right passage. Frequently, on looking over the side, we could see rocks a few feet below the surface on each side of us, and in some parts the passage between them was not twenty feet across. Following the boat, we ran into a narrow channel, between a headland of the main and a string of islands, from which there appeared at first to be no outlet. Suddenly rounding a projecting rock, the boat hitherto our guide let fall her mainsail, and, shoving her bowsprit almost against the door of a small wooden house, her owner stepped ashore with a rope, made fast the boat, and was at home. Before we had recovered our astonishment at this sudden proceeding, we had passed the only place wide enough to swing at anchor, and saw a narrow opening out of the channel between one of the islands and the main. Here was another house, and a man and two boys, who told us the passage was safe, and we slipped through a place where half the crew might have leaped ashore on one side and half on the other. We were then in the small harbour of Port aux Basques, and, by the help of the chart and the sailing directions, worked up to a snug berth behind a small island, where we anchored. Just before it got dark we went ashore ; and found the land very rugged and very difficult to traverse, in consequence of the thick matting of moss and dwarf bushes. There were no houses here, the few inhabitants living at the channel by which we had come in.

CHAPTER VI.

The Dead Islands—George Harvey—Want of Lighthouses
—A Ball—La Poile Bay—Arrival at St. Peter's—The
Cod Fishery — St. Lawrence — Reach Placentia —Visit
Trepassée and Ferryland, and return to St. John's.

September 25th. — A beautiful morning.
Examined the rocks in the neighbourhood of
the harbour, which were nothing but gneiss
and mica slate, the cliffs glittering with crystals and the bright flakes of mica.   In the
middle of the day a light wind sprang up from
the west; so, against Gaden's inclination, we
sailed.   About three p.m. the weather looked
suspicious, and a long sea began to roll in
from the south.   The shore was so rocky and
dangerous, that it was necessary to look out
for a harbour for the night; and seeing a
punt, with a man fishing, near the Dead
Islands, we stood in for him, and he piloted us
through a cluster of small, low, rocky islets,
into a long narrow channel, between a larger

island and the main.   This was a comfortable
and secure harbour, but having shores wild,
bare, and rocky, consisting of gneiss and gra-
nitic veins. Evening rainy and unpleasant ; but
just before sunset it cleared off, and we took the
boat up a small river that came in at the back
of the islands.   For about half a mile it was
several feet deep, with shallow perpendicular
banks of bare rock crowned with moss.   We
then came to a small fall, above which the
brook was a mere rapid, winding through a
bed full of boulders of all sizes and shapes.
It traversed a wild and desolate-looking little
valley, with rocky precipices here and there
on each side of it, and the whole face of the
country was broken into rugged little hillocks
and ravines.   Occasionally, at the foot of some
higher precipice than usual, two or three fir-
trees might be seen trying to struggle into existence.   All the rest was bare rock or lumps of
moss and scrubby straggling bushes. The view
seaward was equally remarkable : numberless
islets ran along the coast, the chief one of which
was half a mile long, steep and woody, but
the majority were mere bare rocks, frequently

in the shape of a low dome, with just one tuft
of bushes growing on their tops.

September 26th.—Fine morning, but the
wind blowing from the south-east right against
us.   There are three or four families scattered
uo the seislets, whom we visited.   They seem
tolerably comfortable, and very simple and
open-hearted, but rather afraid of strangers.
Near the eastern entrance of the channel, in
a house rather better than the others, lived
George Harvey, who had been born and bred
on this coast.   He was about sixty years of
age, and had a large family of sons and
daughters, mostly grown up.   He had been
instrumental in the course of his life in saving
some hundreds of people from wrecks.   A few
years ago the brig Despatch, full of emigrants,
struck, in a heavy gale, on a rock about
three miles from his house.   He and one of
his daughters, then aged seventeen, and a son
of twelve, went off in his punt through a
heavy sea, and by great exertions brought off
the whole of the crew and passengers, consisting of 163 people.   He established the people
about his house, and shared the whole of his

provisions with them, until word could be
sent to La Poile, and a vessel sent down to
carry them away.   They stayed with him a
fortnight or three weeks, and his stores were
so completely exhausted, that himself and his
family were compelled to live chiefly on salt
fish, with neither bread, flour, butter, nor tea,
during the whole of that winter.

On information of his conduct reaching St.
John's, the then Governor, Sir T. Cochrane,
sent him 100*l.* and a gold medal, with directions to wear the latter on holidays, such as
the King's birthday, &c.   This medal, as well
as the Governor's letter, the old gentleman
took great pride in showing us.   On the 14th
September, 1838, he again saved twenty-five
men, the crew of the ship Rankin, 650 tons,
Alexander Mitchell, master, belonging to the
house of Rankin and Gillmore, of Glasgow.
She struck on a rock and went to pieces,
the crew hanging on to an iron bar or rail
that went round the poop.   He fetched them
off by six or eight at a time in his punt
through a heavy surf, and kept them till they
could get on to La Poile.   He showed me
some documents left him by the master of the

vessel, but complained that he had as yet received no recompense either from the master
or the house she belonged to. He and some
more men, he said, were once employed five
days in burying dead bodies cast ashore from
a wreck, from which circumstance, I believe,
arose the name of " The Dead Islands," or,
as the French call them, " Iles aux Morts."
In short, the whole coast between La Poile
and Cape Ray seems to have been at one time
or other strewed with wrecks. Every house
is surrounded with old rigging, spars, masts,
sails, ships' bells, rudders, wheels, and other
matters. A ship's galley lay at Port aux
Basques. The houses too contain telescopes,
compasses, and portions of ship's furniture.
I heard of several chain cables along the
coast, which the people had purchased from
wrecks, on the chance of raising them; and
I believe they drive a regular trade in old
rope and other matters, which are bought by
the little schooners that run during the summer alongshore. Mr. Anthoine afterwards
told me, at La Poile, that there was scarcely a
season (meaning an autumn) but he had four
or five shipwrecked crews thrown on his hands

to maintain and send away, he being an agent of Lloyd's, as well as the only merchant in the neighbourhood.

I should be unwilling to obtrude my opinion on matters of which I have no professional, and but little practical, knowledge;—but surely it seems reasonable that a great commercial nation such as England should not suffer the borders of the great high-road to Canada and her North American possessions to be thus strewed with the property and bodies of her subjects. A lighthouse on Cape Ray, with a large bell or gun to be used in fogs, together with a smaller lighthouse, and a pilot or two, either at Port aux Basques, the Dead Islands, or La Poile, as a harbour of refuge, would be the means of great good. There is no want of good harbours along the coast, but their entrances are generally narrow, and only to be found by those thoroughly acquainted with the channels between the rocks and islands.

Old Harvey told us that the gale which occurred while we were in Codwy Harbour was the heaviest he had known for at least twenty years, and that several small vessels had been lost along the coast. In the afternoon he came

on board to see me, and have a glass of grog;
and then nothing would do but we must all
go back, and have a dance at his house along
with his daughters. Accordingly, we left his
punt alongside our vessel, and took him with us
in our own. One of my men played the flute,
and we got up a rude kind of country-dance,
while he regaled us with grog and tobacco. He
had written a song, which, at his request, one
of his daughters sang to an old " Down, derry-
down" sort of a tune. It was a description of
the wreck of the Despatch, embodying most
of the remarkable incidents that occurred while
he was fetching off the crew and passengers.
The verse, as may be supposed, was rude
enough; but it gave great satisfaction. Then
one of my men, Bill, the flute-player, sang a
song, likewise of his own composition, and de-
scriptive of an adventure of his own, in which
it appeared that he, with the rest of the crew,
had abandoned a water-logged vessel off Ferryland; but, after rowing about all night, were
very happy to get on board of her again the
next morning. This was likewise received
with considerable applause. Old George Harvey then told us of his once having seen a

horse, in some settlement in Fortune Bay,
and described to his family the size and appearance of this remarkable animal. The people
wished, he said, to seduce him into mounting
on its back ; but " He knew better than that,"
although one fellow did ride it up and down
several times. Then, turning to me, " We
have some large beasts, Sir, sometimes, in this
part of the country. I don't know whether
you ever heard of them ;—we call them bears,
Sir." On which I assured him that I had not
only heard of, but had actually seen several
bears in my lifetime, although not at large in
the woods : with all which he seemed especially
interested. I fear the reader will, at first,
hardly feel disposed to believe that there are
British-born subjects, speaking the English
language, in the oldest of our colonial possessions, to whom the horse is a strange animal.
Such, however, is the fact. It was now eleven
o'clock ; and for some time there had been a
storm of thunder and lightning, and heavy
rain and wind, raging outside. Taking advantage of a temporary lull, we set off to return to our vessel. The wind, which had previously blown from the south, shifted into

south-south-west, and began to blow very hard. We accordingly hauled our little vessel close under the lee of the island, and made fast by a line on shore. About two in the morning the wind shifted into the west, blowing a furious gale right down the channel; we were then obliged to haul off from the shore, and let go the other anchor.

September 27th.—The morning broke dark and stormy, with a fierce gale blowing from the west, driving clouds of spray along the water, and heaving up a tremendous sea on the rocks and islands at the western entrance of the channel. The boats alongside were full of water, from the mere "spoon-drift," or spray caught by the wind from the surface of the sea. Luckily, no swell could reach us where we were, and our anchors held on well. About twelve the sun broke through the clouds, but pale and distempered; and it was not till the afternoon that the wind moderated sufficiently to enable a boat to pull against it, even in smooth water. Old Harvey then came off to us, and said that, when he got up in the morning, he ran down to the beach, fully expecting to see us ashore. He was very anxious about

two of his sons, whom he was expecting back
from St. George's Bay, where one of them was
gone to get married.   He fears they may have
been caught in the gale of last night.

September 28th.—The wind has shifted into
the south-east again, not permitting us to stir,
especially in such a heavy sea as was now rolling outside.   A thin, short-haired, black dog,
belonging to George Harvey, came off to us to-
day.   This animal was of a breed very different
from what we understand by the term " Newfoundland dog," in England.   He had a thin
tapering snout, a long thin tail, and rather thin
but powerful legs, with a lank body, the hair
short and smooth.   These are the most abundant
dogs of the country, the long-haired curly dogs
being comparatively rare.   They are by no
means handsome, but are generally more intelligent and useful than the others.   This one
caught his own fish.   He sat on a projecting
rock beneath a fish-flake, or stage, where the
fish are laid to dry, watching the water, which
had a depth of six or eight feet, and the bottom
of which was white with fish-bones.   On throwing a piece of cod-fish into the water, three or
four heavy clumsy-looking fish, called in New-

foundland "sculpins," with great heads and mouths, and many spines about them, and generally about a foot long, would swim in to catch it. These he would "set" attentively, and the moment one turned his broadside to him, he darted down like a fish-hawk, and seldom came up without the fish in his mouth. As he caught them, he carried them - regularly to a place a few yards off, where he laid them down; and they told us that in the summer he would sometimes make a pile of fifty or sixty a day, just at that place. He never attempted to eat them, but seemed to be fishing purely for his own amusement. I watched him for about two hours; and when the fish did not come, I observed he once or twice put his right foot in the water, and paddled it about. This foot was white; and Harvey said he did it to "toll" or entice the fish; but whether it was for that specific reason, or merely a motion of impatience, I could not exactly decide. The whole proceeding struck me as remarkable, more especially as they said he had never been taught anything of the kind.

September 29th.—Sailed to-day with a light wind from the west. The land east of the Dead

Islands preserves the same bleak and rocky character, but gradually rises higher towards the east. About Garia Bay a ridge runs into the country with three high bluffs on it, the high range of Cape Ray being still visible over the intermediate country. We entered La Poile Bay in the afternoon, and anchored in the first cove on the larboard hand, where there is a small settlement, and a considerable mercantile establishment belonging to Mr. Anthoine, a native of Jersey. Mr. Anthoine was in England, and his son very busy; but the latter came on board with Mr. Reid, the custom-house officer, and invited me ashore. There were two brigs and several schooners at anchor.

September 30th and October 1st.—Very unwell with a severe cold and sore throat. Took the boat to the head of the bay, and went nearly round it. It is composed of a coarse-grained granite, of porphyry and quartz, and of chlorite and other slates. The head of the bay is rather more wooded and fertile-looking than the surrounding country, but the eastern shores are craggy and broken, and form a mass of the most irregular, rough, bristling hillocks I ever

saw, rising from a table-land of some 200 or
300 feet in height. I saw here three St. John's
newspapers, being the first news I had had of
the rest of the world for these two months.
We heard also news from the Cape. In the last
gale, that of the 27th, all the vessels lying in
Codwy harbour, consisting of a sloop and two
French schooners, were driven ashore and
wrecked, being beaten entirely to pieces in
about two hours. A large brig was lost, with
all hands, about half a mile above the spot
where we saw the Onondago. She was seen
in the morning to have anchored about a mile
from shore, and inside a shoal on which the
water broke; her crew were clinging to the
yards and rigging. As the wind was dead on
shore, all assistance was, I suppose, out of the
question. At last a sea struck her, and she fell
on her beam ends, but righted again. Another
sea struck her, and she fell over, but not so far.
The next heavy sea, however, turned her clean
over, and she lay bottom upwards. Then the
windlass appears to have been torn from her
deck, and she drifted in-shore, righted as she
approached the land, and was cast into a sandy
cove, where, had the crew run her ashore at

first, they might have been saved. The next day six dead bodies were picked up on the beach, and one was found hanging out of the stern windows of the cabin. Such was the account brought to La Poile; but as nobody who could read had yet seen the wreck, her name was unknown. We were peculiarly lucky in being on the right side of the Cape in each of these gales, and each time in a tolerable harbour. No news had been heard of Harvey's boat, but a boat had been seen coming out of St. George's Bay the evening before the gale, which was thought to be the one in question; and unless they had taken warning in time, and run back up the bay, it was feared they were lost, and that the bridal party had perished in the week of their wedding.

The little settlement of La Poile seemed very comfortable, and the people happy, kind-hearted, and hospitable, although I was too unwell to participate in their enjoyments. Mr. Anthoine's is the chief mercantile establishment between Cape La Hune and Cape Ray. Twenty or thirty years ago there were not more than three or four families within that space. Now, however, there are several inhabitants in almost every cove, and Mr. Anthoine told me that since he came to La Poile he has been obliged to double his stock almost every year in order to supply the increasing population.

October 2nd.—Sailed at daybreak this morning in company with a schooner belonging to Mr. Bagg of La Poile. A lovely day, with a cool breeze. There has been a sharp frost every night for the last few nights. The character of the country continues much the same, but becomes bolder as we proceed to the west. The hills gradually rise higher, but are still bare, craggy, and precipitous; and the country, for the most part, is the very picture of desolation. Mr. Bagg, in his schooner, led the way, and when off the Bruger Islands he suddenly rounded-to, and proposed to run into harbour. The wind had veered to the south, and the sky began to look dirty. We followed him, therefore, through a narrow channel, among many small islands and low sunken rocks, and anchored in a small cove called Grandy's Cove. As we meant, if possible, to sail early the next morning, I landed just at sunset, to get some specimens of the rock, when a man came down,

and, after looking at me with evident wonder
for some time, at last asked me if I wanted
ballast.

October 3rd and 4th.—Detained by contrary
winds blowing fresh from the south-east. Called
on Mr. Cox, who has a small " store" here, and
had married Mr. Bagg's daughter. He gave
us a good dinner of pork, and several kinds of
vegetables, with milk and wine. A regular
dinner was becoming rather an event in my
existence, as I had now for some time lived on
dry ham and rather musty biscuit, my potatoes
being all gone, and no supplies to be got on
this barren coast. Near Mr. Cox's house was
a schooner ashore, half buried in sand: she
was laden with coal from Sidney to St. John's,
and was driven on land in the first gale. There
is a considerable population, inhabiting forty
or fifty houses, which are scattered about the
central portion of the Burger Islands.

October 5th.—Sailed at four in the morning, with the wind from the north. I had intended to have visited Fortune Bay on my return, and also Placentia and St. Mary's. We
had, however, lost so much time, from contrary
winds and other accidents, and it was getting

so late in the season, that I determined on passing Fortune Bay and going direct to St. Peter's, and thence to Placentia. The wind blew fresh from the north-north-east, very cold, and knocked up a troubling sea, that upset my cabin-stove, and so deprived me of a fire. We did not anchor in the outer roads of St. Peter till half-past eight.

October 7th to 10th.—Four most lovely days, perfectly clear, warm, and without a breath of wind. We were accordingly detained in St. Peter's, momentarily expecting a breeze to enable us to proceed. For this detention, to tell the truth, I was not greatly sorry. I was most hospitably entertained by Messrs. Atherton and Thorne, who supplied me with hosts of recent American newspapers and several new books. Two or three of the French officials were also very kind; and M. Renaud, the treasurer, showed me a collection of shells from the Great Bank. Among these shells was a terebratula, which seemed to me of a different species from any I had seen. A new commandant had arrived from France, in place of the one I had formerly made acquaintance with. As, however, I was

in momentary expectation of sailing, I was too
idle to dress and wait on him. St. Peter's was
now very busy, and the inner harbour crowded
with brigs and schooners, the former loading
with fish to take to the West Indies, the latter
chiefly from the neighbouring shores, purchas-
ing French wines, spirits, and groceries, which
they probably intended to smuggle into the
English settlements. Atherton and Thorne's
large store was full all the day. In the outer
roads were several large vessels. One was an
English ship from Miramichi, called the
" Henry," bound for Liverpool. She was dis-
masted in the gale of the night of the 13th,
and put into St. Peter's to refit. Her commander said he never before was out in so
heavy a gale. Indeed I found in the newspapers accounts of wrecks and disasters over
the whole of the western Atlantic from Ber-
muda to Halifax and the Gulf of St Lawrence.
Two English ladies were on board the " Henry,"
one of them with two of her children, and they
said they could never forget the misery of that
night. Many of the inhabitants of St. Peter's
were busy sending off not only their fish, but
themselves, on their return to France. Only

a small proportion of the people engaged in the fishery remain during the winter, the rest return to France; and the French merchants are bound to bring out a certain proportion of fresh hands among their crews every year.   A bounty of about five or six shillings per quintal * is paid by the French government on all fish caught by their own fishermen.   The French fishery is accordingly conducted on a very different plan from the present English fishery. Large vessels with regular crews are employed, every man is numbered, and a regular system adopted, both in the actual catching of the fish and in the curing it.   These vessels fish mostly on the banks, far out of sight of land, and remain there till they have completed their cargo.   Their shore fishery also, on the N.W. coasts of Newfoundland, is conducted on the same regular system, a certain number of boats being under the control of one merchant, and sending in their cargoes regularly to his establishment as the fish are caught, to be there cured.   These boats are, I believe, manned by crews who have regular wages paid them.

* A quintal is 100 pounds weight of dry or cured fish, which fish when fresh would weigh about 300 pounds.

The English fishery is, on the contrary, entirely uncontrolled and unstimulated by boun-
ties.   Every man catches fish for himself or for
others, according to his inclination or his
capital.   He either hires himself out as a servant, for regular wages or a share of the
produce, or he builds himself a boat and
catches what he can, selling the produce to
his merchant, or whomsoever will purchase it.
The English fishery is accordingly now essen-
tially a shore-fishery, and carried on by the
regular inhabitants from their own doors, as it
were.   As might easily be supposed from these
circumstances, occasional squabbles occur on
the shores in the neighbourhood of St. Peter's,
sometimes between the French and English
fishermen, and sometimes between their masters and the authorities.   The French are frequently in want of bait, more especially of
capelin in the capelin season : they are apt,
therefore, to run over to the English shore
with a fast boat and a few nets, and help themselves to a little of it.   This the majority of the
English settlers of course disapprove of, as de-
priving them of their natural advantages : there

are not wanting, however, some of them who
for the sake of profit will undertake to supply
the French with capelin of their own catching,
and this leads to dissensions among themselves. Again, many of the French fishermen
are willing to buy codfish, either cured or un-
cured, from the English, in order to get the
bounty on it from their own government. This
the French authorities naturally oppose, and
enforce stringent regulations against the English fishermen. The English custom-house
officers moreover look on St. Pierre as a nest
of smugglers, or at least as a constant source
of temptation for the English settlers to become smugglers: in which, by the way, they
are not far wrong. On this smuggling-trade
neither the custom-house officers nor the respectable merchants look with favourable eyes.
For all these reasons the presence of a man-of-
war of each nation some time during the
fishing season is absolutely necessary, and
without the salutary fear of such visits the
mutual jealousy would perhaps find vent in
actual outbreaks and open quarrels. As it is,
however, a tolerably good understanding seems

to be kept up, more especially among the
lower orders of each nation, who probably find
their account in it.

October 11th.—After several shifts and
changes, the wind this afternoon settled into a
pleasant breeze from the west. We accordingly set sail, having Mr. R. Thorne, who
lived at St. Lawrence, as a passenger. At ten
o'clock P.M. we were off Cape Chapeau Rouge,
immediately to the east of which are two small
inlets called Great and Little St. Lawrence.
We accordingly stood in, but it was so dark we
could not make out the entrance at first, and
at last got into Great St. Lawrence, instead
of the other inlet, in which was Mr. Thorne's
residence.

October 12th to 16th.—We were again
detained at St. Lawrence during this time by
a succession of contrary winds, during which
I was hospitably entertained by Mr. J. T——
and his lady, who reside at Little St. Lawrence
together with two younger brothers. Their
house is very prettily situated on a small peninsula jutting into the inlet, and connected
with the main by a pebble beach. This beach
is evidently of recent formation, being thrown

up by the indraught of the tide during the
south-west gales, and the peninsula was once
an island. The country around is wild, but
picturesque, having the rocky heights of Cape
Chapeau Rouge in view on one side, and those
of St. Margaret and St. Anne on the other :
these are round-topped craggy hills, each rising
about 800 feet above the level of the sea. The
rest of the country is broken, rocky, and un-
dulated, varied with the usual proportion of
marshes, barrens, and stunted woods, and ornamented with a few fine lakes or ponds, of
which there is one about a mile long near the
north-eastern extremity of the harbour. The
lake empties itself by a stream running between rocky banks, and terminating by a picturesque little waterfall. This waterfall has
cut back about 150 yards in the course of time
through a hard slate rock, forming a ravine
with perpendicular walls, up which the tide
now flows to the foot of the falls. One day
one of the men came across to me from Great
St. Lawrence with intelligence that a lead-
vein had been discovered. On going over I
found quite an excitement, and at least twenty
people hammering and digging at the rock.

The rock itself was a kind of sienite, consisting principally of red feldspar with interspersed crystals of quartz. A vein two or three inches wide was in one part filled with small crystals of fluate of lime, and among these were some little cubical crystals of galena, and a coating of a greenish hue that might be green carbonate of copper. To satisfy the people, I had a good deal of the rock pulled down, and the whole affair shortly ended, the vein closing and losing all its crystalline contents. It is just possible that more mineral veins and of better quality might be found in the neighbourhood, but no other indications existed, and none appeared in the cliffs, while the interior of the country was covered with moss and wood.

October 17th.—Morning fine, calm, and hazy, but towards noon a light breeze sprang up from the west, and we left St. Lawrence for Mortier Bay, where I heard of limestone and marl having been found. The shore is bold, with rocky cliffs and many indentations. We passed close by Burin, in which, from the number of houses, there seems to be a considerable population, and anchored in Mortier

harbour.   Mr. H——, the stipendiary magistrate of Burin, had a house here, and came to
invite me ashore.   The next morning he took
me in his whale-boat into Mortier Inlet, a
short distance farther up the coast.   We passed
through a narrow passage between a small
inaccessible island, called Crony Island, and
the main.   It was scarcely wide enough for
our oars, but a schooner is said once to have
darted through it when chased by an American privateer, and not able to fetch the
principal entrance of the inlet.   There were
several families residing in various parts of this
inlet, in pretty and picturesque situations, and
some spots showed a considerable degree of
fertility.   The different rocks occurring even
in this little space were very numerous,
as may be seen by referring to the report.
There were one or two thin beds of impure
limestone and some red indurated marls, and
" cornstones " in red sandstones, more like the
bottom parts of the coal formation in St.
George's Bay, than anything I had seen on the
eastern side of the island.

October 19th.—Sailed for Placentia, but
about noon the wind headed us, and we made

for the island of Andierne or Oderon. The western side of Placentia Bay is full of islands, many of which are low, and surrounded by sunken rocks and reefs. We had a group of these on each side of us, as we made Cape Jude; and shortly after, passing through a narrow entrance between two small steep islands, we ran into a harbour in another small island, and, narrowly escaping a sunken rock about its centre, we anchored in a snug berth in an inner cove. This was Andierne, and we had scarcely anchored before a north-east wind sprang up, that, had we been outside, would probably have sent us back to Mortier or Burin. On the north side of the little harbour of Andierne is a tolerable house and very extensive warehouses and stores : these are now only half occupied by a young man who was trying in some measure to re-establish the very extensive trade that must once have existed there. One or two brigs were at the wharf, and several small houses were scattered about. We were detained here a day by a north-east wind, and I was hospitably entertained by Mr. H——— on shore.

October 21st.—Sailed with a north-west

wind and a most lovely morning for Placentia.
The air was very sharp and the sky clear, and
as we sailed along I was amused by watching
the rapid changes of form assumed by the
various islands on each side of us as they came
upon the horizon. I concluded—though, having broken my thermometer, I had no means of
testing it—that the sea was of a higher temperature than the air, and that therefore there
was a stratum of air just above the surface of
the sea (which was quite smooth) that continually varied in temperature, and consequently in density and the quantity of moisture
it contained. Such a state of things would, I
should suppose, produce the optical phenomena
alluded to. Although perfectly clear round
about us, the sea near the horizon seemed to
be covered with a dense white fog, smooth and
bright as water, in which the distant islands
were both reflected and distorted. Two or
three headlands were transformed into a row of
narrow pillars, constantly shifting their proportions, and some of the rocky islands assumed all
kinds of queer shapes just before they disappeared. It seemed almost like a narrow stripe
of a phantasmagoria. About three P.M. we ran

into the harbour of Great Placentia. This is
a small bight, with very bold and lofty headlands on the north side, and a low point on the
south side. The highest part of the north
side is covered by the ruins of a fort, called the
Castle, and several old guns are lying about,
one or two having fallen half-way down the
precipitous cliffs. From underneath these lofty
cliffs at the head of the bay, a pebble beach
sweeps out, and is separated by a channel, not
more than 150 yards across, from another
similar pebble beach coming from the opposite side of the bight. On the point of each
of these beaches are the ruins of a strong fort.
The channel leads into the inner harbour, which
divides into two arms : one, called the north-
east arm, about ten miles long, and nearly
straight, and the other the south-east, with a
very winding course of about five miles. The
south-east arm nearly surrounds a steep rocky
piece of ground which was once an island, but
is now connected by a long pebble beach with
the mainland on the south side of the harbour,
blocking up what about sixty years ago was
the entrance of the south-east arm. There is
another pebble beach stretching from it to the

mouth of the channel, on which the town, if it can be called such, is built. It was once a very considerable place, being the French capital when they held possessions on the island, and even under the English was formerly much more important than it is now. It consists of a number of small houses huddled together, and one or two of a better class : it has a Roman Catholic chapel, well attended, and a church, but neither minister nor congregation belonging to it. Mr. B——, the surgeon and magistrate, called on me, and invited me ashore.

October 22nd.—Mr. B—— breakfasted with me, and we then set out to walk to Little Placentia, a distance of about five miles. It was a fine morning with a sharp frost, and to my great delight I found a regular road made. Actually a road! with bridges across the brooks, fifteen feet wide, and mostly covered with gravel. It was rough enough, to be sure, and in some places boggy, and if it had not been for the frost, would have been sufficiently muddy; but I saw none of its inconveniences. I could again step out with a free stride, could raise my eyes from the ground while I walked, and had the free use of my limbs. I never

appreciated the advantage of a road before : it
was really like liberty after imprisonment, and
at the best parts of it I felt very much inclined,
like the Irishman, to walk backwards and forwards in order to make the most of it.   Little
Placentia, as well as Great Placentia, stands on
what was once an island, being connected to
the main by a beach of large pebbles.   It is a
straggling place on the low side of an inlet,
with bold rocky hills on the opposite side.
We borrowed a punt and examined some of
the rocks, which consist principally of slate,
and after being caught in a snow squall, we
dined with a gentleman residing in the town
as a merchant, and returned to Great Placentia
in the evening.   The snow had turned to rain,
and it was blowing hard from the west.

October 23rd, 24th, 25th.—Detained by
westerly winds with heavy rain and miserable
weather.   The entrance of the inner harbour
being so narrow, and the arms so extensive, it
may easily be imagined that the tide, even with
a rise of only six feet, causes very rapid currents at the ebb and flow.   When near low
water, indeed, it is almost impossible to stem the
current of either the ebb tide or the flood with

any strength of wind. The ebb tide is of course the strongest, and also the longest, and the ebb from the south-east arm, in consequence of its many windings, is three-quarters of an hour later than that from the north-east arm : there is consequently a dangerous eddy, and for a short time contrary currents on each bank of the channel. Even with a fair wind, therefore, a vessel is obliged to wait for the tide; and, from the embayed position of the place, one wind is required to take her out of the harbour, and another when she arrives at Cape St. Mary, to take her round to St. John's. At twelve o'clock on the night of the 25th the wind shifted into the north-east, and the tide turning at the same time, we sailed.

October 26th.—It blew very hard last night; and on looking out on my cabin at daylight this morning, I found a pretty scene of confusion. The stove and all its pipes had capsized ; my chair, clothes, and books had got adrift, and together with two or three cups, plates, and other articles, were piled in confusion on the cabin-floor. We were now, however, round Cape St. Mary, with a light

wind at north-north-west. The day turned
out fine, and at noon we were close in with
the east side of St. Mary's Bay, passing
near St. Shots. This place, a small rocky
cove or bay, is the terror of mariners on
this coast. The currents are very variable
and deceitful, and frequently set with deadly
strength right upon this shore, sweeping up
the east side of St. Mary's Bay. Not a season
passes but, at some time of the year, there is a
succession of wrecks just at this spot. Now,
however, it was calm and quiet enough ; and
we sailed past under the steep cliffs of Cape
Freels and Cape Pine, and then ran up into
the harbour of Trepassée. This is a pleasant-
looking place, the land having a gentle slope
on all sides towards the harbour, and being
comparatively bare of wood. There is a considerable number of houses, and among them
one or two of a good size and appearance.
The best belongs to Mr. S——, who, I regretted to find, was from home. He is clerk
of the court to the southern circuit, and was
now absent with the judge, who was on circuit, and whose vessel we saw pass the mouth
of the harbour in the evening. They had left

Placentia shortly before we arrived there, had since been to St. Mary's, and were now gone on to Ferryland.

October 27th.—Rainy and disagreeable, with variable winds from the southward. As the crew had been on deck all the night before, and it was Sunday, I gave them holiday to-day, instead of continuing our route, but had reason most grievously to repent it.

From October 27th to November 9th.—A weary period of thirteen days, the wind blowing from the north-east without intermission. The weather was mostly wet, with fogs, and occasional storms of snow and heavy rain. Even in the fine intervals, I dared not stir far from the harbour, lest a favourable change of wind should occur. I once or twice determined to set out on foot, but the people whom I consulted said the brooks between Trepassée and Renows would not be fordable, except after a day or two of fine weather, and this we never had. I had, moreover, no fancy for wading through several brooks, up to my middle in ice-cold water, if it could be reasonably avoided. Mr. S——'s absence deprived me of all companionship to relieve the tedium

of the time, although Mrs. S—— exercised
all the hospitality she could in the absence of
her husband. Indeed, without her kind assistance, in supplying me with books, and al-
lowing me to join her tea-table in the evening,
I know not how I should have survived this
period of dreary inaction. Of the rest of the
inhabitants, not one came near us.

On November the 2nd the wind had been
blowing very hard all night, and in the morning increased to such a gale as caused our
vessel to drag her anchors, and slowly and de-
liberately drift ashore. Luckily, she settled
on a small beach of pebbles, in the only place
where it was possible to avoid having her bottom knocked in by sharp-pointed rocks. Here,
after a few thumps and heaves, she quietly
settled down among the pebbles; but as the
harbour was two miles long, there was a sufficient swell in the direction of the wind to
dash over her sides, and pour bucketsful of
water into the cabin and hold. Even in this
uncomfortable position, no one from the adjacent houses came to our assistance; and I
was told, I know not how truly, that there
were those on shore who, having made money

by wrecking, cared not how soon we went to pieces, provided it gave them an opportunity of making anything by us. Luckily no ultimate harm was done, and the next day we got the vessel upright, heaved the ballast out of her, and, taking advantage of a high tide on the night of the 5th, floated her off. On the 7th and 8th the harbour gradually got full of vessels, brigs and schooners, waiting for a fair wind for St. John's.

November 9th.—At length the wind came up from the south-west, and we all beat out of the harbour and sailed alongshore with a beautiful day and a fair wind. The land about Cape Race is comparatively low and bare of wood, with a steep cliff of about fifty feet in height, and deep water close in-shore. From Cape Ballard to the northward the land gradually becomes higher to Cape Broyer, a fine bold headland, 400 or 500 feet high. As it got dark the wind became lighter, and off Ferryland it sank to a calm. We accordingly towed into Ferryland harbour, and anchored for the night.

November 10th.—Perfectly calm this morning. Ferryland is a pleasant place, with a

good road along the cliff, and several large and
good houses. The road is continued each way
from the harbour, and towards Aquafort is
sound and in good condition. The scenery
around is varied and picturesque, the bold
headlands, cliffs, and rocky inlets of the sea-
view harmonising well with the dark woods
and hills of the land. At noon, when we were
expecting a breeze to enable us to proceed, the
everlasting north-easter again sprang up right
in our teeth. To avoid a repetition of our Trepassée adventure, with perhaps a worse result,
we got into a sheltered little cove, on the south
side of the harbour, called the " Pool."

November 10th to 15th.—Again did the
wind, during these five days, blow furiously
and pertinaciously from the north-east, bringing snow and frost, rendering it impossible to
stir by sea, and almost impracticable to move
about on shore. I was, however, very kindly
and hospitably treated by several resident families ; and though becoming very impatient,
and being extremely anxious to receive letters
and intelligence, the tedious and wearisome
delay was at length terminated.

On the evening of the 14th, I had just de-

termined to set out to St. John's on foot, when
the wind seemed inclined to change, and on
the 15th a breeze sprang up from the south-
west at twelve o'clock. We accordingly left
Ferryland ; and about eight o'clock at night,
in pouring rain, we had the satisfaction of entering the harbour of St. John's, and I shortly
after landed on one of the slippery wooden
wharfs, and reached at length my lodgings.

CHAPTER VII.

On my arrival at St. John's, I found residing
there Dr. Stuwitz, a Norwegian, Professor of
Natural History in the University of Chris-
tiania. He was travelling at the expense of
his government, to investigate and collect specimens of the natural history of the northern
part of America, and had commenced with
Newfoundland. He intended during the winter to visit St. Pierre, Fortune Bay, and the
southern coast of Newfoundland, and to return
to St. John's in the spring, time enough to accompany the annual sealing expedition, which
leaves that place about the beginning of March.
As I also wished to see the ice-fields which are
visited by the sealers, in order to examine

whether they ever contained boulders of rock
or other matters, I agreed to accompany him.
In the meanwhile I spent the winter in St.
John's, intending to be ready in May to set out
again on my coasting survey, and examine the
rest of the shores of Newfoundland in search of
mineral wealth and productions. During the
latter part of November and the first two-
thirds of December there was dull disagreeable
weather, with occasional snow-storms and frost,
interrupted by thaws. Stuwitz sailed, on the
8th of December, for Fortune Bay, in an open
boat, with a little cuddy at each end, in which
it was just possible to stow a bed, leaving
barely room enough to sit or lie down.

During the latter part of December, and the
whole of January, the weather was beautiful,
the air clear, with sharp frost and snow on the
ground, but no very intense cold. The harbour was never once frozen over, although
the brooks and ponds gradually became fast.
This was the season of general holiday. The
lower orders ceased work; and, during Christmas, they amused themselves by what seemed
the relics of an old English custom, which,
I believe, was imported from the West of Eng-

land, where it still lingers. Men, dressed in
all kinds of fantastic disguises, and some in women's clothes, with gaudy colours and painted
faces, and generally armed with a bladder full
of pebbles tied to a kind of whip, paraded the
streets, playing practical jokes on each other
and on the passers by, performing rude dances,
and soliciting money or grog. They called
themselves Fools and Mummers. The merchants and higher classes shut up their books
and neglected their various employments, and
amused themselves with sleighing parties to
various points where the roads were open; while
a general series of dinner parties commenced,
varied now and then by an evening party and
a dance. There was an amateur theatre, the
profits of which were devoted to charitable
purposes, and a performance took place once a
fortnight, in which their several parts were
well sustained both by the actors and the
audience. There were, moreover, two public
balls, for charitable institutions, that were well
got up and numerously attended. In short,
there was no lack of amusement, till the preparations for the sealing voyage began, towards the middle of February, to draw off the

attention both of masters and men to the more serious business of life.

This winter was remarkably mild; the thermometer in the town never, I believe, sank below zero; and once or twice in the month of February the weather became so warm, that the snow melted on the hills, and a regular thaw took place. The temperature of the air on these occasions could hardly have been less than 50° Fahrenheit, though, as I had lent my last thermometer to Professor Stuwitz, I do not recollect the precise point at which I heard that it stood.*

I cannot, perhaps, find a better opportunity than this of throwing together a few observations on some of the more remarkable facts connected with the trade and productions of the country. The first thing that strikes a stranger on entering a harbour in Newfoundland is the abundance of what are called the fish-flakes and stages, together with the wooden wharfs and the great dark red storehouses.

* Daily observations on the height of the barometer and thermometer have been taken by Mr. Templeman, of the Colonial Office, for seven years, but I regret that I have mislaid his published statement. The lowest degree of cold experienced was, I believe, —18°. The coldest month is February.

The fish-flakes consist of a rude platform, raised on slender crossing poles, ten or twelve feet high, with a matting of sticks and boughs for a floor. On these the fish are laid out to dry, and planks are laid down along them in various directions, to enable the persons who have the care of the fish to traverse them. The surface of the ground being everywhere so rugged and uneven, this device has been resorted to in order to get a sufficiency of flat space ; and in a populous cove or harbour the whole neighbourhood of the houses is surrounded by these flakes, beneath whose umbrageous and odoriferous shade is frequently to be found the only track from one house to the other. The stages are of stronger construction than the flakes, and are generally in the shape of a small pier jutting out into the water, consisting of a platform of poles laid close together, side by side, and nailed down to a strong framework that is supported by stout posts and shores. At the head of the stage are generally two or three poles, nailed horizontally against the upright posts, forming a rude ladder, up which it is necessary to climb from a boat in order to get on the stage. These

are frequently the only landing-places in a
harbour. The central part of the stage is roofed
over, either with boards or boughs, and here it
is that the important operation of splitting and
salting the fish usually takes place. Besides
the flakes and stages, there is generally a set
of rough wooden wharfs, supported on piles,
and floored with boards, at the back of which
are great wooden buildings, some for the re-
ception of cured fish, and others for all kinds of
merchandise. The outside of these buildings
is painted according to the taste and fancy of
the owner, but usually of a dark dull red. Now
if the reader will picture to himself, in addition
to all this, a few brigs and schooners at anchor in
the harbour, and a multitude of small fishing-
boats, varying in size from a two-oared punt
to a half-decked schooner-rigged craft of ten
or fifteen tons;—a broken rocky shore, with a
stunted wood and little patches of cleared and
cultivated garden-ground;— one or two large-
sized wooden houses, painted white, belonging to the merchants, and a number of unpainted wooden cottages scattered here and
there at all possible angles with each other,
perched upon rocks and hidden in nooks, be-

longing to the fishermen :—if, I say, he can associate all these things in his fancy, he will have before him a tolerably correct notion of a Newfoundland settlement. Of course, in the larger places, more especially in St. John's, these elements are overpowered and thrown into the background by others of more imposing character. Large stone houses, good-sized churches, chapels, and court-houses : shops built of wood and painted white, a tolerably regular street, and a road or two, mark the seat of greater wealth, and a more numerous population. Even in St. John's, however, fish-flakes are by no means entirely absent, though they are confined to the south side of the harbour, and to a small nook, bearing the euphonious appellation of Maggoty Cove.

About the beginning of May all the population are on the alert, preparing for the fishery, laying in stores of summer provisions, and hooks, lines, nets, clothes, and the rigging of their boats. Towards the middle or end of May the first shoal of herrings, called by the natives the spring herrings, appear. These are immediately caught in nets and used as bait for the cod fish. In the middle of June the

capelin come in and last to the middle of
July; and with them commences the height
of the fishery. Every man, woman, and
child is then fully employed. A married man
having a family usually goes out with his
sons, takes his bucket full of capelin for
bait, and rowing to the fishing-ground, generally a mile or two outside the harbour, anchors and commences fishing. Each person in
the boat has two lines about twenty-five fathoms long, with two or more hooks. These
he flings one on each side of the boat, the end
of the line being made fast to the thwart.
Feeling each occasionally, the moment he
strikes a fish, he hauls him in, flings him
down in the boat, baits the hook, and throws
out his line again. When they get what they
call a good spurt of fish, each person will
sometimes be fully employed hauling in one
line after the other, as fast as he can bait
them and throw them out again. When
this happens, an hour or two suffices to fill
the boat, which then sails away with her
cargo to the stage-head. Here the fish are
forked out of the boat with a kind of boat-
hook or pikel, the prong being stuck into the

head, and the fish thrown up on to the stage
much in the same manner as hay is thrown
into a cart. On the stage are usually the females of the family, or a man or two assisted
by females, as the case may be. The two most
skilful of these are called respectively the header
and the splitter. The business of the first is to
cut the fish open across the throat and down
the belly, and pass it to an assistant, who taking
out the liver drops that into a tub on one side,
and tearing off the head and entrails throws
them down on the other side. The liver is preserved to make oil, and the head and garbage
drop into the water which flows underneath.
The fish is then passed to the splitter, who by
a dexterous movement cuts out the backbone
from the neck nearly to the tail, and thus lays
the fish entirely open and capable of being laid
flat on its back. This is the most important
part of the operation, and a goods plitter always
commands superior wages at a merchant's establishment. When split open, the fish are
salted, laid in piles to drain, washed and salted
again, and finally laid in the sun on clear
days to dry and harden. While thus exposed
they require much attention, and the women

are constantly looking after them, laying them
up in round heaps with the skin outwards at
night or on the approach of rain, in which state
they look very much like small haycocks.
Towards the end of July and the beginning of
August, the capelin leave the shores, and the
young squids or small cuttle-fish succeed them
in myriads and supply their place. These are
caught and cut up for bait, being a very favourite
food of the cod ; and when their season is over
these are succeeded by the " fall herrings," as
they are called, or the autumnal herring-shoals.
This is in September, which may be looked on
as the close of the fishery. During the whole
season the shell-fish, both of fresh and salt-
water, are an excellent bait for the cod ; and
sometimes food is in such enormous abundance, that the fish get gorged and refuse all
baits. In such case a jigger is resorted to, an
implement I have already described as a plummet of lead armed with hooks, and drawn
quickly up and down in the water, attracting
the fish by its motion, and striking them as
they swim around it. The jigger, however,
is looked upon as objectionable, and ought
only to be employed in cases of necessity, as it

wounds more fish than it catches. I do not
know the amount ever caught by a man with a
hook and line in a single day, but it must sometimes be enormous, as it is stated by them as a
cause for their devoting their whole time to it
that they *have the chance* every day of catching
5*l.* worth of fish. Now cod-fish at present is
not more than 15*s.* a quintal when dry and
ready for exportation, and a quintal of dry fish
is made from about three hundred weight of
fresh or " green " fish, consequently 5*l.* worth
of fresh fish would be very nearly a ton weight.
To get this quantity a man must catch 224
cod of an average weight of 10 lbs. each, or a
greater number of less weight in a day; and
this, from what I have seen, I think by no
means an impossible occurrence, and I have no
doubt it sometimes takes place. A family of
five or six active individuals in a good summer
may thus make fish of the value of from 50*l.*
to 100*l.* currency. Young and single men
either join together, or hire themselves out at
regular wages or for a share of the produce,
either to their own neighbours or to the merchants, the wages being generally 20*l.* for
the summer, with rations. Lastly, many fa-

milies in some of the outports, instead of
"*making*," or curing, their own fish, bring it
as it is caught to the merchants' stores and
stages, where it is cured by his own men.

In some parts of the coast, where the water
is sufficiently shallow for the purpose, the
cod-fish are now caught in seines or other
nets. This operation requires more capital
to commence with than the mere boat and
hooks and lines of the common fishermen.
It is therefore chiefly pursued by the merchant, or by the richer and more considerable
of the " planters," and a great jealousy exists
on the subject in some places. Some people
even go so far as to say that all nets should
be prohibited, as destroying the chance of
the poorer class. Setting aside the difficulty, however, of such a prohibition, there
are some places, as about Greenspond and
Cape Freels, where the net is used almost
exclusively; and little cod would be caught
without it. It is obvious, moreover, that the
use of the net is advantageous to the trade
at large, as shoals, or, as they are called,
" schools" of fish, may sometimes be seen
sweeping alongshore that refuse to bite at

all; and, but for the net, would escape altogether. Besides there seems such an incalculable abundance of the fish, that there will always be enough to hook, enough to jigg, enough to net, and more than enough to go away. One calm July evening I was in a boat just outside St. John's harbour, when the sea was pretty still, and the fish were "breaching," as it is termed. For several miles around us the calm sea was alive with fish. They were sporting on the surface of the water, flirting their tails occasionally into the air, and as far as could be seen the water was rippled and broken by their movements. Looking down into its clear depths, cod-fish under cod-fish, of all sizes, appeared swimming about as if in sport. Some boats were fishing, but not a bite could they get, the fish being already gorged with food. I speared one great fellow with the spike of the boathook; but there being no tail to it, he got away; and, as far as I could see, that was the only fish touched. Had the ground been shallow enough to use nets, the harbour might have been filled with fish.

The nature of the connection between the

merchants and the settlers, fishermen, or "planters," as they are called, is peculiar, and is
now somewhat complicated. When the fishery
was first embarked in by the English, it was
conducted much as the French fishery is now.
Vessels of a large size came out every year for
the summer, and were compelled to bring the
whole of their crews home in the winter, and
take out a certain proportion of fresh hands
every year. At that time the Bank fishery
was the most important branch ; and the
fishery alongshore little attended to. Gradually the country became thinly settled along
the coast, and at length was recognised as a
colony, and a governor appointed. Still the
connection between the fishermen and the
merchants continued much as before. The
merchants were men of large capital, and each
had his regular planters, who were either his
hired and bound servants, or had been such,
and were entirely dependent on him for supplies. The large mercantile establishments,
then in comparatively few hands, were scattered along the coast ; each establishment had
a regular crew "shipped," and bound down
by articles of agreement ; and the neighbour-

ing planters, whether bound or not, had no
other market than the next merchant's store,
either to take their fish to, or procure provisions and goods from. Every man in the
neighbourhood was down in the merchant's
books for almost everything he had; little
hard cash passed, but each account was balanced by the fish or labour of every man
during the season. As, however, the merchant
fixed the price of the fish he bought, and also
of the goods he sold, there was considerable
probability of the ultimate result being in his
favour. From the great facility, moreover, of
procuring credit with his merchant for anything he needed or wished for, the careless
fisherman frequently ran in debt to a greater
amount than his summer's work would pay
off. In winter he must either get provisions
on credit or starve, and his debt must be
increased before he could commence his summer's work to clear it off. The idle and dis-
honest ran away, or refused to pay: this was
a loss to the merchant, who, to cover the
losses from bad debts, put a higher price on
his goods; and thus the good were obliged

to pay for the bad. With such a system, even with honest intentions on both sides, the natural consequence must be, that each merchant or agent held his neighbourhood in a species of vassalage, and had more power over his dependants than it is safe to trust to the will or caprice of any man. The settlers, on the other hand, became careless and improvident: they lost the feeling of independence, and that never-failing spur to exertion, the hope of acquiring it. This system, too, involved great temptations to dishonesty on both sides; and if the open evasion of a debt on the one side was the most flagrant, I am not sure whether the silent overcharge on the other was not the most fraudulent and pernicious. The close of the war, and consequent fall of the price of fish, led to the breaking up of some of the large mercantile establishments; others failed from various circumstances; the increasing and more stable population drew people with smaller capital to set up stores in a smaller way, and opened the door to competition; and the larger houses concentrated their business in St. John's, or a few of the principal places, and supplied the

merchants in the out-ports, or any persons
who would pay for their goods, either in cash,
fish, or oil.   Lastly, the number of small ped-
dling schooners trading along the coast, frequently stepping in between the merchant and
his planter, and buying the fish from under
his nose, as it were, acting in concert with the
other causes, gradually broke up the old
system, while political and religious differ-
ences completed the alienation between the
fishermen and the merchant.   Most of these
circumstances were in themselves evils, and
all have been bitterly complained of; but as,
on many occasions, great and general good
arises out of many and partial evils, these
changes also will, I believe, ultimately be productive of good, and establish the trade of the
country on a far more stable and healthier
system than before.   In the neighbourhood of
St. John's, and the whole of Avalon, the old system may be said to have almost entirely passed
away.   The fisherman, if honest and industrious, may carry his fish to any one he
chooses; and though he cannot fix the price
at which it shall be sold, as the merchants

fix that by common consent from the state
of the foreign markets, he has still the great
benefit of competition in the choice of the
provisions and goods he is to buy. When
once he has acquired a boat, a house, fishing-
lines, and provisions for the season, he has
the power of being independent, and need
not owe a farthing.   One or two years' in-
dustry, backed by a little good luck, will
give him these ; and five years' work, at most,
ought to place him on the road to honest
independence.   Annihilate, as far as possible,
the system of credit, and educate the rising
generation (which is now being done), and the
people of Newfoundland may be as happy
and prosperous a race as any under the sun.
The value of the dry cod-fish alone exported
every year from Newfoundland is, on an average, about 400,000*l.*, while the total value
of the exported productions in fish, oil, and
skins is upwards of 700,000*l.*   This, from a
population of 70,000 or 80,000, proves the
extent and value of their resources, especially
when we take into consideration the quantity
of fish consumed in the country.

To resume, however, with the usual course
of labour of a planter and his family. In
some places, they remove at the close of the
fishing season, with their families and furniture, from their summer residences, into more
sheltered and woody districts, where fire-wood
is more abundant, and the materials for making oars, punts, staves, hoops, &c. may be had
for the labour of cutting them. Others employ
themselves in hunting, in shooting wild-fowl,
ptarmigan, which they call partridges, or deer,
or in trapping martens, foxes, otters, and other
animals, for their fur. The latter employment, however, except in very remote settlements, is seldom resorted to; and its place
is supplied in Avalon and the north coast by
the sealing expeditions. In some favoured
spots, namely, at the mouths of the principal
brooks, salmon-fishing is followed during the
summer, by one, two, or more families; and
in all places, besides the cod-fish, the dog-fish
is caught for the oil extracted from its liver;
and the herring and capelin are sometimes
cured, as well as used for bait. There is a
whaling establishment in Fortune Bay; and

I have often wondered, from the abundance of whales and grampus on the northern shore, and more especially in Trinity and Bonavista Bays, that no whaling speculation has been there set on foot.

For the statistical details of Newfoundland, I must refer to Mr. Montgomery Martin's work on the statistics of the colonies, and to a little volume in the Edinburgh Cabinet Library, in which a very fair account is given of Newfoundland; not entirely accurate in minor points, but, upon the whole, good.

It is an ungrateful task to turn critic on the characters of a people among whom one has lived, and of whose hospitality one has partaken. I can, however, assert that the Newfoundlanders are simple, honest, industrious, goodnatured and hospitable people, and have the virtues of all hardy races exposed to the toils and dangers of an adventurous life. Some of them are, no doubt, fond of rum; but though, during Christmas, or in holiday times, they may occasionally be " fou for weeks thegither," the mass of the

people are not habitual drunkards.   Many
are tee-totallers of old standing.   The Roman
Catholic part of the population, more especially, have an old custom called "cagging,"
taking a vow, that is, before the priest, not to
touch rum or spirits for a year or two, or for
their whole lives, or while they are on shore,
&c., and these vows are scrupulously adhered
to.   Of late years, excited by elections, and
the instigations of a few turbulent spirits and
by the old party feelings and enmities imported
from Ireland, there have been one or two
outbreaks at St. John's and Carbonear; and
some acts of private or public vengeance have
been perpetrated, to the extent of maiming the
person.   Scenes quite as bad have taken place
at elections at home, more especially, I believe, in Ireland; but they naturally produce
more excitement in a scattered and usually
tranquil population.

One point in the character of most of
the inhabitants of Newfoundland seems to be
common to the whole of North America,
namely, the eager inquiry after news, and
the propensity to exaggeration and invention,
to use the most polite terms.   It is really

astonishing how news of the most trifling
matters, especially if dashed with a little scandal, flies about, not only in St. John's, but
along the coast to the most distant settlements.
Reports of the most ludicrous nature,—ludicrous for exaggeration, even if they have any
foundation in fact,—gain instant credence. It
seems to be a stain on a man's character if on
coming into a harbour he has not a budget of
news; so that if he knows none, he immediately draws on his imagination. The seal
hunters, and the furriers in the country, make
a point of giving false information as to the
results of their expeditions; and they were,
once or twice, quite angry with me for telling
the truth on these points. "Sure, sir, what's
the use of letting them know what we've got?"
"But what's the use of telling a lie about it?"
"Faix! and it's no lie; what call have they
to be asking about it?" The consequence of
this indifference to truth is a bad one. Mali-
cious sayings, and tale-bearings, reports of
private conversations, and remarks with ill-
natured emphasis or additions, and all the
petty malice of scandal, are rife in all the
settlements I visited; often introducing the

most bitter private dissensions into communities that might otherwise be happy and united. The first society in St. John's is imitated in this matter by the humble fishermen in the most retired out-harbour, in a manner that to the unprejudiced observer seems the most severe satire upon them.

The point most deficient, however, in the character of most of the lower classes of the inhabitants is a want of manly independence and self-reliance. They are easily led, and always look for guidance; being ready to follow any one who will take the trouble of thinking and deciding for them. This feeling has resulted, in the English part of the population, from their habit of depending upon and looking up to their merchant for everything; and the same effect has been produced amongst the Irish from the implicit reliance they place upon their priest. It is evident that, where this feeling of dependence exists, there can only be a negative character, for without some degree of self-reliance and self-decision, a man may do little wrong, but neither is he likely to do any positive good. I be-

VOL. I.   M

lieve, however, that these pernicious influences are on the decline; and that mercantile competition and early education will gradually elevate the character of the people of Newfoundland.

I have already alluded to their propensity to take advantage of the calamities of their neighbours, during fires and shipwrecks. They universally reason in this way: "If I don't save these goods, they will be lost: now I may as well have them as the sea or the fire!" My own men I found constantly talking in this way when speaking of a wreck; and certainly, when a vessel, entirely abandoned, is driven ashore, and neither her name nor owners are known, the reasoning seems fair enough. It is probably in this way that a lax feeling has originated; and, when once prevalent, it is not unlikely to lead to various infractions of the rights of property. But this feeling is already on the decline in the more populous parts of the island; and the noble examples which the merchants and inhabitants of St. John's have afforded to the people at large cannot but have their effect. While

I was in the country, there were several instances of shipwrecked crews, and large bodies
of emigrant passengers, not merely clothed,
fed, and taken care of, but refitted, their losses
partly compensated, and themselves forwarded
free of expense to ports near their places of
destination. In proportion to its size and
wealth there are few places in any part of the
world where larger and more frequent sub-
scriptions for charitable purposes are from
time to time raised than in St. John's.

Of the political state of the country I shall
forbear to speak at any length. There is
now, unhappily, very considerable bitterness of
party spirit; but what the cause may be no
one seems able to tell. There are no political
principles involved in the disputes : indeed, I
cannot call to mind having ever once heard a
political principle stated, either publicly or
privately, while I was in the country. The
old nicknames of Tory and Whig are bandied
about, and there are dissensions between the
Roman Catholics and the Protestants; but the
sole feelings involved are, it would appear,
certain conventional prejudices, and the strug-

gle for personal or party power, influence, and emolument. The whole population is British, both actually and in thought and feeling : they are all loyal subjects of the English Crown ; and the idea of separation from the mother country has never entered into the head of any, even the most violent among them. On the whole, I believe that if the points of difference could once be fairly understood and expressed, and if the two parties could agree among themselves as to the distribution of the public offices, their quarrels would die away simply for want of anything further to dispute about. Their political arrangements, therefore, can hardly be of sufficient general interest to call for further notice in this place.

It would be hardly fair, perhaps, to class fogs among the productions of this country ; but political obscurities lead naturally in this place to speak of the fogs for which Newfoundland is so well known. These fogs are certainly most remarkable. They prevail generally during the summer months, June, July, and August. They do not originate on shore, but at sea, and more especially in that part of

it which washes over the Great Bank of Newfoundland. The Gulf-stream, sweeping along
the eastern coast of North America, strikes off,
about the latitude of New York, towards the
north-east, and reaches the southern edge of
the Great Bank; but the boundaries of the
stream are indefinite, and it advances farther
towards the north at one time than at another.
From Baffin's Bay to Davis's Straits a great
current runs towards the south, sweeping along
the coast of Labradore. This current per-
plexes and renders dangerous the navigation
of the Straits of Belle Isle, whence another
current flows out of the Gulf of St. Lawrence,
and after washing the north-eastern coast of
Newfoundland, traverses the northern edge of
the Great Bank, varying in power and extent
according to circumstances. The tract of sea
between these two currents and near their
junction must, of course, be subjected to considerable eddies, changing in power, extent,
and direction, according to variations in the
two currents. Such eddies render the sea
south of Newfoundland and between that country and Nova Scotia dangerous to the navi-

gator, as he is uncertain in what direction, and at what rate, he may at any time be drifted by them.   It is doubtless to the meeting of these two currents, one charged with ice, and the other warm from the tropics, on the Great Bank,* and the shoals about it, that the fogs are chiefly due.   On the southern shore of Newfoundland a dense white fog prevails throughout the summer, whenever the wind does not blow from the north or west. This fog is low, and does not reach far into the country, as frequently on ascending a hill two or three hundred feet high the sun is found shining brightly overhead.   Even on the low ground small breaks or open patches sometimes disclose the bright sun for a few minutes, but it is soon again shrouded by the dense white vapour.   These fogs are not unhealthy, nor very unpleasant to the senses, and they rarely extend so far as the Gulf of St. Lawrence, St. George's Bay, or the western shore of Newfoundland.   All along the southern

---

* The Great Bank has a depth varying from sixty to fifteen fathoms.   Forty fathoms may with great probability be given as its mean depth.

shore they are very prevalent, more especially in Placentia Bay, and the neighbourhood of St. Pierre; but around St. John's, on the eastern side of the island, they may frequently be seen out at sea, while the land is basking in sunshine. Sometimes a strip of sea, about a mile broad, forms a belt around the shore, within which the fog does not penetrate, so that vessels suddenly emerging from the fog have found themselves in a clear atmosphere within a mile of the shore. The appearance of a vessel coming thus through a wall of fog is described as most singular, her jib and then her foresails being seen to emerge before her mainmast comes into sight. I have often, too, been struck with the appearance of the narrows when a light easterly wind has been blowing. Overhead, in the harbour, and on the landward slope of the south-side hills the sun has been shining brilliantly, while in the gorge of the narrows the fog might be seen boiling and rolling in huge white fleecy folds, creeping along the water, or curling in wreaths of mist up the sides of the hills, and gradually dissipating as it advanced. Sometimes the crest of the ridge of the south-side hill was crowned by a long low line of fog that kept rolling over it and pouring down upon its sunny sides without advancing an inch, the whole of the vapour being dissolved as soon as it came upon the warm air above the heated rocks. If, however, the wind blew harder from the east, or more especially the south-east, the fog was driven in more rapidly: it gradually advanced and prevailed, spreading over the harbour, and enveloping the whole country in its dense mantle. At night, when the ground is cold, the fog settles upon the land; and remains as long as the wind blows from the east or south-east. The moment, however, the wind shifts into the west, or north, the fog is driven out to sea again, and retires to its stronghold and permanent domain, the Great Bank. With northerly or westerly, or at St. John's with south-westerly winds, the atmosphere is always clear; but during the summer months, a shift to the south or east will bring fog in greater or less time, according to the strength of the wind. In Bonavista Bay, and along the northern shore of Newfoundland, there is comparatively little fog, the only wind that brings it being a north-easterly, or a very strong easterly wind. I have no doubt the interior of the country of Newfoundland would always be found free from fogs, except in long-continued southerly and easterly winds.

CHAPTER VIII.

Return of Dr. Stuwitz, and departure with him to the Ice—
Appearance of the Ice, and difficulty of getting through
it—Capture of the first Seal—Grand slaughter amongst
the Seals—Icebergs—Aurora Borealis—Capture of a
Shark — Description of the different species of Seal
found near Newfoundland—Effects of a Swell upon the
Ice—Return and arrival at St. John's.

TOWARDS the end of the month of February,
1840, I began anxiously to expect the return
of Professor Stuwitz, of whom I had heard no
intelligence during the winter. In seasons of
ordinary severity his return by sea to St.
John's would have been impossible; and he
intended to have made for the nearest port
that was free from ice, and have come on
overland. The mildness of the weather,
hitherto, had prevented any great accumu-
lation of ice along the coast, although the
harbour was frequently skinned over, and its

mouth alternately opened and blocked up by
ice and frozen snow on each influx and reflux
of the tide. Two or three severe days would
have been sufficient to close it entirely, or
a field of ice from the northward might drift
down and beset the whole coast. However,
on the 25th, I had the satisfaction of seeing
his boat making her way into the harbour,
and he soon landed on the wharf. He had
suffered greatly, having frequently been out
at sea all night in his open boat in the severest
weather: he had, however, made many good
observations, and collected specimens of the
land and sea fauna both of St. Peter's and
Fortune Bay. There was now but little time
left to make arrangements for our voyage to
the ice. The generality of the vessels " going
to the ice" are schooners and brigs from 80 to
150 tons, manned by a stout crew of rough
fishermen, with a skipper at their head of their
own stamp. These ships are inconceivably
filthy, being often saturated with oil; and crew
and skipper all live and lie together in a narrow
dark cabin of the smallest possible dimensions,
and the fewest possible conveniences, everything being in common. In such a vessel as

this it was evidently impossible for us to be accommodated, nor would either crew or skipper have consented to take us. There are, however, one or two masters of ships of a superior description who reserve the after-cabin to themselves, and keep the crew in the forecastle. Among these, one of the most respectable was Mr. Furneaux, who having been at one time unsuccessful as a merchant, had retrieved his fortune by successful sealing enterprises; and he had just built a new vessel called the "Topaz," a brigantine of 120 tons, in which he was going out for the first time. She had a comfortable little after-cabin, with small state-rooms, containing altogether five berths. This cabin he agreed to share with us; Professor Stuwitz and his baggage were to occupy one state-room, myself and servant another, while the larger and after state-room was reserved for Captain Furneaux and his more important stores of powder, spirits, &c. I had discharged Kelly, and engaged a young fellow named Simon Grant, a clean, active, good-tempered lad, who was to act as cabin-steward and cook. Our expedition was looked upon in St. John's as rather a Quixotic undertaking, no

one, except the men actually engaged, having
yet had the curiosity " to go to the ice ;" and
these men either could not give clear descriptions, or purposely made rather a mystery of
the matter.   At length, on

Tuesday, March 3rd,—We had got everything on board, and were all ready for a start.
The wind was in the morning unfavourable,
but began to get variable towards the middle
of the day, and we were anxiously looking for
a change.   Being heartily tired of waiting,
however, and Captain Furneaux saying he
should not start for an hour or two, at least,
I strolled with the gentleman at whose wharf
the vessel lay into the billiard-room.   We had
finished three rubbers, when, happening to look
out of the back window, we saw over the roofs
of the houses the topsails of a vessel loose, with
the house-flag of my companion at the mast-
head : throwing down our cues, we rushed
into the street, and the first person we met
informed us that messengers were searching
for us all over the town.   I found on the
wharf-head a host of friends assembled to take
leave, the vessel with her sails loose and only
moored by a single rope, and a punt alongside

waiting for me. I jumped into the punt, sculled off, clambered up the side of the vessel as they cast off the rope, and before I could get my foot on the deck we were under sail. Three cheers greeted us from the wharf as our sails filled, and the crew were called aft to return it: one cheer more from the wharf-head, and we were off. The harbour was covered with thin ice, and frozen snow, that yielded before the pressure of the vessel; and, as we slowly advanced, cheers resounded from several others of the wharfs as their respective vessels cast off their moorings and set sail. We passed two smaller vessels which were stuck fast in a pan of ice, and endeavouring to make their way out; another brig joined us as we entered the narrows, in company with which we sailed from the harbour and lay alongshore to the northward. In consequence of my billiard playing I had lost my dinner, but we had hardly gone a mile before the keen air recalled this fact to my recollection, and I dived into the cabin to examine the interior of a meat pie. This was about four P.M., and yet at half-past seven I was at supper with the rest, with a hearty appetite, on hot beef-steaks. Our vessel, the Topaz, seemed to sail very well on a wind, as we drew a-head of all the vessels we came near. The swell was very trifling, and there were scarcely any waves. The surface of the water was in a half-congealed state, and had a thick greasy appearance, as if it were of the consistence of pea-soup: this was owing either to a quantity of snow resting on the water, or to the first formation of small spiculæ of ice, and it gave a very peculiar appearance to the long lazy swell of the sea. At sunset we were off Cape St. Francis, with a long line of white on the north-eastern horizon, showing the edge of the ice to be in that direction. During the night, while fast asleep in my berth, I was awakened by a loud rushing and roaring noise, with continued vibration and heavy jars now and then, which 1 found to proceed from the ship having entered the ice, the thin sheets of which were scraping and rubbing against her sides, while here and there a heavy lump shook her as she struck it. At last we stuck fast, and the crew were then called to free her, and she was put about, sailed back to the southward, and kept more in-shore in a broad channel between the ice and the land.

March 4th.—At daylight this morning we were just at the entrance of Trinity Bay, and should have been much farther but for the detention of last night. A broad belt of clear water stretched across Trinity Bay, from Baccalieu Island to Cape Bonavista, while ice occupied the interior of the Bay on the one hand, and stretched off far away to the northward on the other. Many vessels were in sight, some in the clear water, others in the ice. Near two of the latter, which were in the ice five or six miles to the north of us, we could just discern some little black dots slowly moving about: these were part of the crew breaking the ice, and trying to cut their vessels out. Much of the ice near us was in the state the men call "lolly," by which they mean soft, half-frozen snow, floating on the surface of the water, not more than five or six inches in thickness, and yielding readily to pressure. On reaching Cape Bonavista we found the channel along which we had hitherto advanced blocked up, and the ice close to the shore, with only a few open pools here and there, and small channels leading to nothing. Several vessels were tacking to and fro along the edge of the ice, waiting for a favourable

opportunity of making their way through it.
Taking advantage, however, of a part where
the ice was loose and broken, we pushed
through, doubled the Cape, and then followed
a small channel, in the direction of two or
three vessels ahead. In a mile or two this
likewise closed, and for the rest of the evening
we continued sailing about in such pools of
smooth water as were left in the intervals
of the ice. The aspect of this ice was very different from that of fresh-water ice, being much
softer, quite white, and opaque, and looking
rather like frozen snow than ice. It was
generally about a foot thick, but was cracked
in every possible direction, and broken into
large cakes or pans, of all shapes and sizes,
that were sometimes jammed and crushed
together, and sometimes slowly moved on and
made to revolve by the grinding of their edges
together. Frequently one piece was tilted up
and rested partly on another, and here and
there small pieces were lying loose on the surface of others. The sides and edges of the
pans seemed frequently water-worn into holes
and hollows, and sharp spurs sometimes pro-
jected under water. Some of these masses of

ice of a smaller size had been washed and worn on all sides by the waves into the most irregular and fantastic shapes, now resembling the turrets and battlements of an ancient castle, and presently shifting and looking like the picturesque cliffs of a rocky shore. It was a lovely evening, mild, and pleasant on deck, the thermometer not having been below 40° Fahr. during the day. South of us lay the headland of Cape Bonavista, from which the varied shores of the Bay stretched away to the north-west, the surface of the land being patched and· streaked with snow. Seaward was one vast expanse of white ice, with small lakes of dark water dotting its surface here and there, but the dreariness of this view was relieved by the number of vessels in sight.

This evening the crew were divided into watches and punts-crews, and prepared their bats and gaffs, expressions whose meaning will be better understood by the following description of the method adopted in manning a sealing vessel. The vessel undertakes to carry a certain number of men, and she is fitted out, provisioned, and found in everything by her owner, or a merchant hiring her, as the case may be. The men then apply for a berth on board of her, and all those who are selected *pay so much for their berth*, generally about four pounds currency, and for this they are supplied with the ships' provisions during the voyage. The results of the voyage on their return are divided into two equal halves, one of which is shared among the men, the other divided between the captain and merchant in different proportions, according to their agreement or their various shares in the risk of the speculation. Captain Furneaux was both master and owner of the Topaz, and I believe he bore all the risk of the voyage, paying for her supplies and engaging his crew. The crew consisted of thirty-six men, who were divided into three watches, each under an appointed master of the watch, specifically engaged under that designation, and receiving something more than the rest of the crew. The men had each equal authority, and each acted as mate during his own watch, the charge of the deck being left to his care when the captain was below. We carried nine four-oared punts, three of them commanded by the three masters of the watch, and the other six by men

picked from the crew, and likewise appointed
by the captain. These punt-masters then drew
lots, each one putting some article, such as a
knife, &c., into a cap or bag, which was
brought to the captain, who drew the articles
out at random, and as he drew them out their
owners were arranged accordingly. Then, in
the order in which they were thus arranged,
each in succession selected a punt out of the
nine on deck, and afterwards chose the men to
make up his crew. Each punt had, therefore,
three men and a master attached to it, who
alone were suffered to man it; and in cases of
emergency, by calling out the number of the
punt or the name of its master, it was ready
and manned, without confusion or delay. This
having been settled every man prepared his
"gaff," by firmly fastening a spiked hook like
a boat-hook, with strong line, to the head of a
stout pole, about six or eight feet long; several yards of strong cord were also selected,
which were prepared with a noose, for a
"hauling rope," the use of which will be seen
hereafter.

By the time that all these arrangements
were made night was falling; and running

the vessel's nose into a stout pan of ice alongside, we brailed up and furled our sails, and lay
secure and motionless as if moored to a wharf.

March 5th.—This morning was dark and
foggy, with the wind at south-east. At seven
o'clock, after making a tack or two about an
open lake, and finding no channel, we dashed
into the ice, with all sail set, in company with
two other vessels, on a north north-west course.
The ice soon got firmer, thicker, and heavier,
and we shortly stuck fast.   " Overboard with
you! gaffs and pokers!" sung out the captain;
and over went, accordingly, the major part of
the crew to the ice.   The pokers were large
poles of light wood, six or eight inches in circumference, and twelve or fifteen feet long:
pounding with these, or hewing the ice with
axes, the men would split the pans near the
bows of the vessel, and then, inserting the ends
of the pokers, use them as large levers, lifting
up one side of the broken piece and depressing
the other, and several getting round with their
gaffs, they shoved it, by main force, under the
adjoining ice. Smashing, breaking, and pounding the smaller pieces in the course the vessel
wished to take, room was afforded for the mo-

tion of the larger pans. Laying out great claws on the ice ahead, when the wind was light the crew warped the vessel on. If a large and strong pan was met with, the ice-saw was got out. Sometimes, a crowd of men clinging round the ship's bows, and holding on to the bights of ropes suspended there for the purpose, would jump and dance on the ice, bending and breaking it with their weight, shoving it below the vessel, and dragging her on over it with all their force. Up to their knees in water, as one piece after another sank below the cutwater, they still held on, hurraing at every fresh start she made, dancing, jumping, pushing, shoving, hauling, hewing, sawing, till every soul on board was roused into excited exertion. After looking on some time, I could stand it no longer : so, seizing a gaff, I jumped overboard, but soon got a damper, as, in my first essay to cross before the vessel, I did not distinguish the sound pieces from the mere broken mash and lolly, and in I went to my middle before I was aware of it. One of the men caught hold of me, and I scrambled up the side of the vessel, a little cooler than I went down. Every fresh hand, they said, has to pay

his footing for his first dip : so I was obliged
not only to lose my footing in water myself,
but give it afterwards in rum to the crew.
They continued their exertions the whole of
the day, relieved occasionally by small open
pools of water; and in the evening we calculated we had made about fifteen miles.

It continued foggy all day, and at night it
began to rain. We had seen no vessel since
the morning—nothing but a dreary expanse of
ice and snow stretching away into the misty
horizon.

March 6th.—At daylight we found ourselves
a few miles from Green's Pond, on the north
side of Bonavista Bay, and with the glass could
discern the houses and the church. The wind
was from the west, and the sky fine and clear.
Several vessels were near us, and many more
on the horizon. The ice became thicker,
stronger, and more compact. We made a few
miles in the morning, and stuck fast the rest
of the day in a very large pan or field of ice,
sawing, axing, prising, warping, &c. &c., as
yesterday. The thermometer during the day
was about 37°, and the weather pleasant. I
walked about on the ice to get accustomed to

it. In the afternoon a long swell came in from
the eastward, raising the fields of ice in long
slow undulations as far as we could see. There
was, however, no open water visible in any direction from the mast-head. In the evening
it was mild, with a drizzling rain.

March 7th.—Fresh breeze from west and
north-west, cold but fine. A vessel close by
left St. John's just after we did; her captain
came on board and made a morning call. The
ice gets continually heavier as we proceed to
the northward; great blocks and slabs of it are
tilted up and piled on one another here and
there, as if occasionally jammed and forced together with great violence. In the afternoon
the ice opened a little, the pans became more
loose, and separated a little one from the other.
We accordingly crowded all sail, and cracked
on with a fresh breeze, urging our way now
through slowly yielding pans, now sailing gal-
lantly through an open lake of water, and then
crashing into a great sheet of ice, splitting it
with a mighty fissure, but brought up ourselves all standing and top-gallants set. Then,
slowly gathering way again, we proceeded as
before. The motion while in the ice is like

that of a steamer, continually trembling and
vibrating with the thumping and rubbing of
the ice, while the scraping of the pieces against
the sides of the vessel makes a noise like the
roar of machinery. The weather was fine and
clear, and in the afternoon we got one or two
good runs through some large lakes of water,
and hove to during the night in a smaller one,
the wind blowing a heavy breeze.

   March 8th.—Quite calm this morning, with
large pieces of ice all around us. I shot a sea-
bird, called in Newfoundland a mur, which is,
I believe, some species of mergulus; and saw
also a large snowy owl flying about, but could
not get a shot. I found also on a piece of ice
the greater portion of one of the common
echini of Newfoundland, broken, and full of
dried mud. This must have been inclosed in-
the ice originally on the beach, in some bay
or inlet, as, if brought by a bird, it would have
contained no dirt. We were now, however,
at least forty miles from land, and surrounded
by ice on all sides. In the course of the morning I spied a great seal on the ice about a
quarter of a mile off, and, taking my rifle and
a man with a gaff, went to get a shot. We

found, however, the ice in that direction broken
by many channels full of "lolly," which were
too wide to leap. In returning, I slipped in
jumping from one pan to another, and fell into
some of this " lolly :" luckily I had passed the
gun to the man, and he had given me the gaff,
which, as I fell, was caught by its ends upon
the ice, and by this means I got across to my
companion, who hauled me out. I had heavy
seal-skin boots on, coming half way up the
thigh, which being now full of lolly formed a
delightful case for my leg had I wished it
frozen. However, the ducking seemed to do
me good, and, oddly enough, quite cured a
cold which I had felt coming on before the
accident. In the evening a cold north wind
sprang up, and the ice closed in on every side,
locking us quite fast. Thermometer on deck
at six P.M. 24°.

Last night, as we sailed rapidly through the
broken ice and water, there was an abundance
of the phosphorescent flashes and sparks so
well known at sea. Many of these sparks
were observed, not only in the water, but either
in or upon the small lumps of ice. They must
probably have been in the water which washed

over the ice, but they had every appearance of being *in* the ice, whose interior appeared to be lit up by them. They were most beautiful objects, and might have formed fitting lamps for the caves of the Nereids. Stuwitz found infusoria in ice in Fortune Bay, which, when the ice was melted, were living and uninjured.

March 9th.—Light wind from north. Thermometer at nine A.M. 22°, and in the afternoon 27°. We passed this morning through some very heavy ice. It was broken and piled in huge slabs and blocks, resting one upon another in various directions, the cracks and crevices filled up with snow, and frequently worn by successive thaws and frosts, rains, winds, and waves, into the most striking and fantastic forms. This broken ice occupied a rather definite band for some distance, and appeared to have been once the external edge of the ice on which a heavy sea had beat violently. Many of these blocks were higher than the rail of the vessel, or about fifteen or twenty feet from the surface of the sea, and as many yards in diameter, with upright sheets and pinnacles rising occasionally still higher.

March 10th.—At dawn a gale from the south brought rain and dirty weather, and we lay all day fast in the ice with our sails furled. In the afternoon we filled all the water-casks that had up to this time been emptied with beautifully clear fresh water, collected in the hollow of a pan alongside of us. After dark the wind fell, and the clouds gradually cleared off before a light westerly breeze, unfolding a most lovely sky studded with bright stars, and adorned by the presence of the young moon, and the brilliant flickering streamers of a fine aurora in the north. The ice too opened, and we sailed gently through calm water, among numberless fairy islets of glittering ice and wreaths of snow, with shining pinnacles and fantastic forms floating calmly about us, and—

" Quietly shining to the quiet moon."

Everything was still, and even the sailors hushed their noisy clamour, insensibly silenced by the influence of this most lovely scene. Even the hoarse voice of the master of the watch as he sung out, from the foretop, brief orders to the helmsman, was not out of har-

mony with the feelings of the time, while,
sounding at intervals, it served but to make
the silence of all nature around us more deep
and solemn. For a long time Stuwitz and
I stood silent in the bows of the vessel, en-
tranced by the novelty and beauty of all
around and above us. It seemed the realiza-
tion of the poetic visions of early youth in
its dreams of the fleeting and unearthly loveli-
ness of fairy-land. Gradually, however, the
aurora faded from the sky, the vessel passed
from the open pools of water into a field of
closely packed ice-pans; and the enchantment
was dissolved by the cry of " Gaffs and pokers!"
and the usual thumping, grinding, and shouting of our passage through close ice.

March 11th.—Light breeze from the S.E.,
with fog and hoar frost. Thermometer at 10
A.M. 33° F. The vessel making but little way,
the ice being closely packed. Two men came on
board this morning from a vessel at no great
distance, but which we could not see for the
fog. It was singular, in such apparently utter
solitude and desolation, to see two human
beings making their appearance. After a chat
with their friends on board the Topaz, and a

glass of grog, they went off again into the fog
to return to their vessel. The watch on deck
last night heard the cry of a young seal. The
weather remained the same all day, but in
the afternoon the ice opened, and we got a fair
run of some distance to the northward.

March 12th.—Again foggy, with a south-
east wind. As we stood on deck this morning
before breakfast, we heard a cry down to lee-
ward, like the cry of a gull, which some of
the men said it was. It became, however, so
loud and continued, that both Stuwitz and I
doubted its being the cry of any bird, and one
of the men took a gaff and went to look. We
watched him for some distance with our glasses
as he proceeded slowly through the fog till he
suddenly began to run, and then struck at
something, and presently returned dragging
a young seal alive over the ice, and brought
it on deck. It was of a dirty white colour,
with short close fur, large dark expressive eyes,
and it paddled and walloped about the deck
fierce and bawling. A Newfoundland dog called
Nestor, belonging to the Captain, approached
it, but it snapped at his nose and bit him,
though its teeth were but just beginning to

appear. After taking it down below to show
the Captain and demand the usual quart of
rum for the first man who caught a seal, one
of the men knocked it on the head and skinned
it. Stuwitz then cut off its " fippers" or paws
and its head, and after breakfast we took it
into the "after-hatch," or steerage, where he
drew and dissected it. In the middle of the
day we heard from some of the men who had
been out on the ice, that a vessel a few miles
a-head of us had already 2500 seals on board,
so we pushed on through the ice, and shortly
came into a lake of water. On the borders of
this many young seals were lying, and two
or three punts were hoisted out to despatch
and collect them. I shot one through the
head that was scuffling off a pan of ice, but
the crew begged me to desist, as they said
the balls might glance from the ice and
injure some of the men who were about.
Having picked up the few which were immediately about us, we hoisted in our punts again,
as there were several vessels near us, and more
coming up, and bore away farther north
through an open pool of water. In passing
through a thin skirt of ice, one of the men

hooked up a young seal with his gaff. Its
cries were precisely like those of a young child
in the extremity of agony and distress, something between shrieks and convulsive sobbings.
It thrilled one's nerves at first, but when I
found that their sole employment, when alone
on the ice, was uttering similar cries, and that
nearly the same cry was expressive of enjoy-
ment or defiance as of pain and fear, I became
more reconciled. We soon afterwards passed
through some loose ice on which the young seals
were scattered, and nearly all hands were overboard slaying, skinning, and hauling. We then
got into another lake of water and sent out
five punts. The crews of these joined those
already on the ice, and dragging either the
whole seals or their " pelts " to the edge of the
water, collected them in the punts, and when
one of these was full brought them on board.
The cook of the vessel, and my man Simon,
with the Captain and myself, managed the
vessel, circumnavigating the lake and picking
up the boats as they put off one after another
from the edge of the ice. In this way, when it
became too dark to do any more, we found we
had got three hundred seals on board, and

the deck was one great shambles. When piled in a heap together the young seals looked like so many lambs, and when occasionally, from out of the bloody and dirty mass of carcasses, one poor wretch still alive would lift up its face and begin to flounder about, I could stand it no longer; and arming myself with a hand-spike I proceeded to knock on the head and put out of their misery all in whom I saw signs of life. After dark we left the lake and got jammed in a field of ice with the wind blowing strong from the north-west. The watch was employed in skinning those seals which were brought on board whole, and throwing away the carcass. In skinning, a cut is made through the fat to the flesh, a thickness generally of about three inches, along the whole length of the belly from the throat to the tail. The legs or "fippers," and also the head, are then drawn out from the inside, and the skin is laid out flat and entire, with the layer of fat or blubber firmly adhering to it, and the skin in this state is called the "pelt," and sometimes the "sculp." It is generally about three feet long and two feet and a half wide, and weighs from thirty to fifty pounds.

The carcass when turned out of its warm covering is light and slim, and, except such parts as are preserved for eating, is thrown away.

March 13th.—As soon as it was light this morning all hands were overboard on the ice, and the whole of the day was employed in slaughtering young seals in all directions and hauling their pelts to the vessel. The day was clear and cold, with a strong north-west wind blowing, and occasionally the vessel made good way through the ice, the men following her and clearing off the seals on each side as we went along. The young seals lie dispersed here and there on the ice, basking in the sun, and often sheltered by the rough blocks and piles of ice, covered with snow. Six or eight may sometimes be seen within a space of twenty yards square. The men, armed with a gaff and a hauling rope slung over their shoulders, disperse about on the ice, and whenever they find a seal strike it a heavy blow on the head, which either stuns the animal or kills it outright. Having killed or at least stunned all they see within a short distance, they skin, or, as they call it, "sculp" them with a broad clasp-knife, called a sculping-

knife, and nicking two holes along the edge
of each side of the skin, they lay them one
over another, passing the rope through the
nose of each pelt and lacing it through the
side holes, in such a manner that when pulled
tight it draws them into a compact bundle.
Fastening the gaff in this bundle, they then
put the rope over the shoulder, and haul it
away over the ice to the vessel. In this way
they bring in bundles of pelts three, six, or
even seven at a time, and sometimes from
a distance of two miles. Six pelts, however, is reckoned a very heavy load to drag
over the rough and broken ice, leaping from
pan to pan, and they generally contrive to
keep two or three together to assist each other
at the bad places, or to pull those out who fall
into the water. The ice to-day was in places
very slippery and in others broken and treacherous, and as I had not got my boots properly fitted with " sparables " and " chisels " I
stayed on board and helped the captain and the
cook in managing the vessel and whipping in
the pelts as they were brought alongside. By
twelve o'clock, however, my arms were aching
with this work, and on the lee side of the vessel

we stood more than knee deep in warm seal
skins, all blood and fat. Some of the men
brought in as many as sixty each in the course
of the day, and by night the decks were covered, in many places the full height of the rail.
As the men came on board they occasionally
snatched a hasty moment to drink a bowl of
tea, or eat a piece of biscuit and butter; and as
the sweat was dripping from their faces, and
their hands and bodies were reeking with blood
and fat, and they often spread the butter with
their thumbs, and wiped their faces with the
backs of their hands, they took both the liquids
and the solids mingled with blood. The deck
of course, when the deck could be seen, was
almost as slippery with gore as if it had been
ice. Still there was a bustle and excitement
in the scene that did not permit the fancy to
dwell on the disagreeables, and after a hearty
refreshment the men would snatch up their
gaffs and hauling ropes, and hurry off in
search of new victims: besides, every pelt was
worth a dollar! During this time hundreds
of old seals were popping up their heads in
the small lakes of water and holes among the
ice, anxiously looking for their young. Occa-

sionally one would hurry across a pan in
search of the snow-white darling she had left,
and which she could not recognise in the
bloody and broken carcass, stripped of its
warm covering, that alone remained of it.
I fired several times at these old ones in the
afternoon with my rifle from the deck, but
without success, as unless the ball hits them
on the head, it is a great chance whether it
touch any vital part, the body being so thickly
clothed with fat. In the evening, however,
Captain Furneaux went out on the ice and
killed two with his sealing-gun loaded with
seal shot. The wind had now sunk to a light
air, and the sun set most gloriously, glancing
from the golden west across the bright expanse
of snow, now stained with many a bloody
spot and the ensanguined trail which marked
the footsteps of the intruders on the peaceful-
ness of the scene. Several vessels came up
near us from the south, in the afternoon; but
notwithstanding all the slaughter the air as
night closed in resounded with the cries of
the young seals on every side of us. As the
sunlight faded in the west, the quiet moon
looked down from the zenith, and a brilliant

arch of aurora crossed the heavens nearly
from east to west, in a long waving line of
glancing light, slowly moving backwards and
forwards from north to south across the face
of the moon. The evenings after a north-west
wind are certainly most lovely, the atmosphere clear and transparent, and having that
dry crispness and elasticity that makes every
breath send the blood dancing with fresh vi-
gour from the heart. It is on a clear night,
moreover, beneath the milder light of the
moon, and contrasted with the deeper blue
of the sky, that ice scenery looks best. In
daylight it is too dazzling, garish and mo-
notonous : the moon and the stars, and the
quivering aurora, are its fittest accompani-
ments.

March 14th. — Wind north-east, with fog
and snow. Early in the morning the crew
were out on the ice, and brought in 350 seals.
The number hauled in yesterday was 1380,
making the total number now on board upwards of 2,000. After suffering the pelts to
lie open on deck a few hours, in order to get
cool, they are stowed away in the hold, being
laid one over the other in pairs, each pair having

the hair outwards. The hold is divided by
stout partitions into several compartments or
" pounds " to prevent too much motion among
the seal-skins and keep each in its place.
The ballast is heaved entirely out as the pelts
are stowed away, and the cargo is trusted to
to ballast the vessel. In consequence of
neglecting to divide the hold into pounds in
one of his earlier voyages, Captain Furneaux
told us he once lost his vessel. He was detained on his return, with 5000 seals on board,
by strong contrary gales which kept him at
sea till by the continual motion and friction
his seals began to run into oil. The skins
then dashed about from one side of the hold to
the other, with every roll of the vessel, and he
was obliged to run before the wind, which was
then blowing from the north-west. The oil
spread from the hold into the cabin and forecastle, floating over everything and forcing the
crew to remain on the deck. They got up
some bags of bread, and by putting a pump
down through the oil into the water-casks they
managed to get fresh water. After being in
this state some days, himself and his crew
were taken out of the vessel by a ship they

luckily fell in with, and carried to St. John's, New Brunswick; but his own vessel, with her once valuable cargo, and almost all the property of both himself and his crew, were necessarily abandoned to the mercy of the winds and waves, and what became of her was never known. This was a good practical lesson as to the proper method of stowing a cargo of seals, and one not likely to be forgotten: in the present instance, therefore, the pounds were both numerous and strong. In the middle of the day the wind increased to a gale, with a heavy drift of snow, making it impossible to stir far from the vessel, and we shortly made a start and sailed with a close-reefed fore topsail through broken ice and pools of water. Suddenly we fell in with a number of seals, and in the hurry of the moment I followed several of the crew overboard to the ice, with my gun, and made for an old one I saw not far off. All the surface being quite white, I took it for granted we were fast in a great field of ice, and only just discovered, time enough to prevent my stepping into it, that the surface of the water was covered with snow which congealed as it fell, and that there were only a few

pans of ice around the vessel, which was still
under way. By this time the Captain and
Stuwitz were shouting to me to come on board,
and running to the bows of the vessel across
the last pan of ice that touched her, I had only
just time to pass up my gun to a man that
stooped over for it, and swing into the bight of
a rope, before we were in clear water, driving
furiously along. My essays hitherto were
thus eminently unsuccessful. The Captain
now took in all sail, and we scudded under
bare poles amid clouds of snow, till we stuck
fast in a large field of ice. We were then securely harboured, the men turned in, fastened
down the hatches, and we in the cabin ate,
drank, smoked, and read the few books we had
with us, making ourselves as comfortable as
circumstances would permit. Our only drawback of any consequence was the smokiness
of the stove, but I never was in a vessel yet
in which the smoke did not come down the
funnel of the stove at least as often as it went
up it.

March 15th.—I was awakened last night by
the vessel rocking, for the first time since we
had entered the ice, and found that the field

about us had opened into great bays of water.
It was now blowing a furious gale, and we were
sailing under the spencer through loose ice and
open pools. The decks were as slippery as
glass with frozen snow and blood ; the thermometer being down at 17° at two P. M. We
still picked up a young seal or two in the
morning, but in the afternoon were again fast
among close ice, and there was nothing more
to be done but to repeat the amusements and
occupations of yesterday. Towards evening
the wind moderated a little, but was still very
cold.

March 16th. — Beautiful morning, with a
light wind from the west. Sailing through
thin ice and open spaces of water. There were
many vessels in sight, and several icebergs,
some of the latter of considerable size. At ten
A. M. the thermometer was 27°. We did not
get near any icebergs, but looking at them
with the assistance of the telescope I could see
nothing but ice and snow. A large one about
three miles from us had apparently been half
tilted over, there being a perfectly straight indented mark across it, which had evidently
once been its water line, though now it crossed

the mass diagonally at an angle of about 45°
with the horizon. Above this water line the
ice was rough and jagged, below it smooth,
though not level, and evidently water-worn.
Behind the iceberg appeared a large low
mass that may have been broken off from
it, and thus caused it to tilt over. The field
ice appeared to get thinner as we proceeded
northwards, many pans, apparently firm, being nothing but frozen snow, through which
one could drive a gaff, and one or two of
the men got a ducking in consequence. I
went out in a punt, and at a distance of not
more than twenty yards positively drove a
ball into the throat of an old seal that rose
to look at us; notwithstanding which, he dived
and got away. We brought one young seal
on board to-day, alive and unhurt, by my par-
ticular request. He lay very quiet on deck,
opening and closing his curious nostrils (which,
when expanded, were nearly round, but closed
firmly into a narrow slit), and occasionally lifting his fine dark lustrous eyes as if with
wonder at the strange scene around him. His
fur being quite dry and clean was as white
as wool—short, and close, and thick, com-

posed of strong hair standing out perpendicularly everywhere, except. on his face and his fippers, or paws. On being patted on the head he drew it in, making his face perpendicular to his body, knitted his brows, shut up his eyes and nostrils, and he then presented a very droll appearance, having a comical countenance in a circular bush-wig. With his head extended, however, and his eyes open, he was really a very pretty creature, looking so warm, and round, and comfortable. When teased, although quite young, he was fierce, biting and scratching at everything about him; but on being patted and stroked he immediately became quiet. These animals might certainly be easily tamed if properly managed. The men and the dog, however, teased this one till he became exhausted and dying, and I passed my knife into his heart, to put a stop to it at once. In the afternoon two of the men went out on the ice with their sealing guns, and shot three bedlamers, or seals of one year old, and brought them on board, where Stuwitz measured, drew, and described them. The sealing-gun is an immense affair, as long as a duck-gun, but with a much wider bore, roughly made, and, in some instances, not over-sound. The men put in a great charge of powder and shot—frequently ten fingers' breadth, or even more—the powder being coarse, and the shot larger than buck-shot, consisting, in fact, rather of small bullets than shot, being cast, and not dropped. It was as much as I could do to hold one of these guns straight out; and the men were frequently knocked down by its rebound, when they fired standing on slippery ice.

At twelve o'clock to-day we took an observation, for the first time, and found our latitude to be 50° 17'. What our longitude was we were obliged to guess—no chronometer could stand the jarring and thumping it would receive in so small a vessel among the ice; and as for keeping a dead reckoning, that was entirely out of the question. Working and winding our way among the ice wherever there seemed the best opening between the pans, changing our course every minute, the vessel brought with her head up in the wind by the pressure of a pan of ice at one minute, turned off by another right before it the next minute, sailing now rapidly through a pool of water, then stuck fast for half an hour in the ice, and, even when apparently stationary, drifting in an unknown direction, at an unknown rate, with the whole body of the ice—the log is of course utterly useless. All we knew was that we were north or north-east of Newfoundland, in latitude 50° 17′. This is the parallel of the south coast of Devonshire, where the winters are so mild, and where a lump of ice out at sea would astonish the inhabitants not a little. What would these good people think if the English Channel, the Irish Sea, and the German Ocean were blocked up with icefloes, and if it were possible to cross from France to England, and thence to Ireland; to proceed northwards and pass over first to Scotland, and then to Norway, and afterwards coming southwards, to return again from France to England, and all on solid ice? Had the western coast of Europe the same climate as the eastern coast of America, this would be an occurrence within the limits of possibility, and certainly not a winter would pass without all the seas round England being blocked up with ice; our early crops would then perish; our woods would become stunted, and our

oaks, ashes, and elms give way to pines and
larches: mosses and lichens would spread over
our corn-fields and pastures, and England
would become a desolate and barren country.

In the afternoon a St. John's schooner came
alongside, with scarcely any seals on board,
and followed us for the rest of the day, hoping
to profit by our good fortune. Captain Furneaux was a noted man as a lucky seal-hunter;
the luck, in fact, consisting in energy and
perseverance, and never stopping or losing a
minute till he did find them. In the evening
the wind fell to a light air, and the sunset was
beautiful. We were sailing quietly to the
north, through thin, broken ice, in hopes of
meeting with some of a thicker and firmer
description, on which there was a greater probability of finding seals. The moon was
bright overhead, with thin streaks of cloud
in patches about the sky, dimly veiling the
stars; and though it was cold, we stood long
on deck watching the shadow of the sails as
we passed successively from the white ice to
the dark water.

March 17th.—A beautiful morning, quite
calm and mild. In the afternoon a light air

sprung up from the south-east, and it became
rather foggy; the ice was very loose, and the
pans, covered with snow, were very rotten and
treacherous. The decks were now cleared up,
and everything put snug and comfortable again.
In some of the open pools of water to-day I
observed a singular appearance. Spaces fifty
or sixty yards across were covered with a thin
film of clear ice, that seemed to be made up of
round scales the size of one's hand, frozen
together into larger flakes or scales about a
yard in diameter : these frequently overlapped
each other, and being very bright and shining,
looked like the scaly side of a gigantic fish.
We gathered some snow from the surface of a
pan to-day that seemed quite salt, the spray of
the sea having probably been blown over it.

March 18th.—Fine morning, light breeze
from the east, the vessel making some way
among ice very loose and rather thin. The
farther we go north the more thin and broken
the ice seems to get, much of it seeming to be
merely frozen snow, which, alighting on the
surface of the sea, has become saturated with
water and then frozen hard. We got one
young seal again this morning, and an old

hooded seal made his appearance alongside, being the first we had seen—all we had hitherto met with being harp-seals. At eleven o'clock we came within two miles of a schooner that was moored to a pan of ice on one side of a lake of open water, and we sent a punt accordingly to see if there were seals in the neighbourhood. As soon as they touched the ice, we saw the men commence killing and hauling, so we immediately hoisted out all the punts and went after them. I went with one punt, taking my double-barrelled gun and some seal-shot. I got several shots at old seals, but found the charges from my fowling-piece not heavy enough for them, as, though I wounded several severely, most of them got away, and I only killed one or two dead on the spot. Even the young ones, unless hit about the head or in the heart, will carry off a good deal of shot. I knocked over one young one that was shuffling off a pan, but notwithstanding this he popped into the water, and swimming about ten yards, crawled on to another pan, when I gave him the other barrel: he again got into the water, and on his crawling out one of the men fired a sealing-gun at him,

but the contents only striking him about the
tail he again got away, and it was only by
shooting him in the head as he raised it from
the water, that I succeeded in killing him at
last.   Generally, however, the young ones
did not attempt to stir as we approached them,
and quietly suffered themselves to be knocked
on the head with the gaff, and skinned on the
spot.   I saw one poor wretch skinned, or
sculped, while yet alive, and the body writhing
in blood after being stripped of its pelt.   The
man told me he had seen them swim away in
that state, and that if the first blow did not
kill them, they could not stop to give them a
second.   How is it one can steel one's mind to
look on that which to read of, or even to think
of afterwards, makes one shudder?   In the
bustle, hurry, and excitement, these things pass
as a matter of course and as if necessary, but
they are most horrible, and will not admit of
an attempt at palliation.   As this morning I
was left alone to take care of the punt while
the men were on the ice, the mass of dying carcasses piled in the boat around me each writhing, gasping, and spouting blood into the air
nearly made me sick. Seeking relief in action,

I drove the sharp point of the gaff into the brain of every one in which I could see a sign of life. The vision of one poor wretch writhing its snow-white woolly body with its head bathed in blood, through which it was vainly endeavouring to see and breathe, really haunted my dreams. Notwithstanding all this the excitement of being out in the punt, forcing our way through narrow channels between the ice-pans into lakes of water where old seals were sporting on every side, filling our boat with pelts, engineering so as to sweep over a mile or two of new ground, and return to the vessel by a different way from that we came, clearing the ice of seals as we went along, and the hunting spirit which makes almost every man an animal of prey, and delight in the produce of his gun or his bow, kept me in the punt till a late hour in the afternoon. In our first trip we brought thirty-three on board; in the second, forty-four; and in the third, a still greater number. Thirty old or " round seals," or sixty pelts, are as many as the boat will carry with four men. One boat, however, brought one hundred and three at one trip, having upwards of forty in tow in the water,

but she was down nearly to the water's edge.
Stuwitz made, I believe, several interesting
observations to-day on both old and young as
they were brought on board, as also on the
still-born young, which the men call cats, and
of which we found several lying on the ice.
By night-fall we again had our decks covered with seals' pelts, and the men were
busy all night heaving out ballast and stowing
them away.

March 19th.—Foggy, and nearly calm: a
few punts were out in the morning, but it was
too thick to go far, and there were but few
seals near us. There were 1080 brought in
yesterday, and about 100 this morning, making
about 3300 now on board altogether. We
sailed about an open lake of water all day, and
at night went into the ice. A heavy swell
was now rolling in from the east, raising all
the ice into long sweeping undulations of
considerable height and breadth : there was
not, however, much drifting or clashing of
the ice-pans together, but each piece rose and
fell perpendicularly on the waves that successively travelled beneath it, without other apparent motion. The thermometer for the last

few days has generally risen to 37° or 40° in the course of the day.

March 20th.—Very thick and foggy, with light wind from the north-east: two or three vessels near us, but no seals. We breakfasted this morning on the hearts and kidneys of young seals fried :—they were very good, being just like pig's fry, but rather more tender and delicate. At ten o'clock we pushed into the ice on a north-west course; presently we heard some young seals " bawling " a-head, and in a short time we got among them, and most of crew went after them. The captain shot an old hooded seal on the ice, and Stuwitz and I went to cut him up and examine him. He was upwards of six feet long, and four or five in circumference in the thickest part of his body: on his nose and forehead was a great pulpy mass, like a black bag full of fat, over-hanging above his eyes. This is the hood which, when angry or excited, can be blown out to a considerable size. He had a cold, malignant-looking eye, and his head altogether looked fierce and *outré*, something like a hip-popotamus or rhinoceros. Stuwitz occupied about an hour in measuring and drawing him, after which we sculped him and cut him up. The flesh was very black, and strong, and muscular, the fibres of the muscles still quivering when we cut into them, although he had been dead so long. His bones, on the contrary, were very thin and slender for so bulky an animal. We were another hour in dissecting him, up to our elbows in black blood; and as it froze on our hands and arms we were obliged occasionally to plunge them into his body to thaw and warm them.

The fat was between two and three inches thick; the skin, when it got dry, was a very beautiful one, being a dark silver-grey, elegantly spotted with black: this is the natural colour of all the seal skins, although when brought home the grey is changed to a dirty yellow, being soaked in the hold of the vessel with rancid fat. The fields of ice this morning were thrown into such strong undulations by the swell, that at a distance of fifty or sixty yards a man seemed to be alternately on the top of a sloping bank and in a hollow, by which half his person was hidden from sight. When walking and looking around, it gave a feeling of insecurity and giddiness, but when standing or kneeling, and looking down, not the smallest degree of motion was perceptible. I tried this two or three times, and after looking steadily down for a short time, was always obliged to raise my eyes to convince myself that the swell still continued. In the afternoon the fog thickened while some of our men were out a long way after seals; the captain and I accordingly amused ourselves with firing a nine-pound carronade that was in the bows of the vessel once or twice every half hour as a signal. On one of these occasions, not having raised the breech enough, we blew away the part which was raised over the muzzle of the gun, and a piece of it flew backwards over my head, with which it very nearly came in contact. Before dark two men came alongside, looking for eight men missing from their vessel, and shortly afterwards two more from another vessel, who had lost their way in the fog. Gradually all our crew found their way back, and the four strangers slept on board of us. Throughout the night we could hear guns and signals of various kinds from different vessels around us

whose men were missing. Beating the outside
of the bulwarks with a rope makes a noise that
is heard a long way; clapping two boards
together, or striking a frying-pan with a poker,
is also a good signal, the sound being more
distinct, and the quarter it proceeds from more
easily distinguished than the heavy boom of a
gun through the fog.

March 21st.—Thick, foggy, and calm. The
four men went off into the fog in search of their
respective vessels. We scarcely stirred all day,
and frequently heard guns and signals for lost
men, some of whom must have been out on
the ice all night. This thick fog is one of the
principal causes of danger in a sealing voyage,
men having often been lost, being drifted out
to sea on the ice, separated perhaps from their
vessels by channels suddenly forming, or wan-
dering quite away in the fog, and perishing
miserably of cold and hunger on the surface
of a frozen sea. We picked up a few young
seals, and as one of them was shedding his
white coat, we eased him entirely of it, and
disclosed his second coat, a beautifully spotted
skin of short smooth hair—grey and black.
The weather felt quite warm, the thermometer

being above 40°, and we took off the cabin skylight.   Ever since we left St. John's we have
had flies in the cabin, but now the skylight
was full of them, buzzing about till they were
quite a nuisance : some were even flying about
the deck this morning.   After dark we heard
a great noise, as of breakers ahead, which we
supposed was an iceberg, with the sea beating
against it, and therefore, for fear of accidents,
we moored the vessel fast to a large thick pan
of ice alongside, on which were several upright
blocks and masses.

March 22nd.—A beautiful morning, with
the wind at north-north-west, quite clear.
There were many vessels in sight, ten or a
dozen of which had flags flying, signals that
they had lost men.   One small schooner passed
close by us with six men short, whom she could
not hear of, but hoped they were on board
some vessel or other.   We discovered the noise
we heard last night to proceed from what is
called a " rolling " pan.   This was a small iceberg of very irregular shape, about sixty yards
across, and its average height above the water
about twenty-five feet.   It drifted down upon
us by the aid of an under current, as there was

but little wind, and obliged us to cast off our
tow-line and make sail. It was full of hollows at the top and sides, having a pinnacle
at one end thirty-five feet high; with a perpendicular wall or cliff, and at the other a round
hummock; it consisted of white compact and
opaque ice, enclosing at one place a large wedge-
shaped mass of blue transparent ice. But what
made it most remarkable was its swinging to
and fro in the water, dipping first on one side
and then the other into the sea, and heaving
up tons of water as it rose, that, surging into
its hollows, fell back down several channels
into the sea, producing a noise and motion exactly like that of breakers. There was but
little swell, but what there was sufficed, I suppose, to keep up the rolling, which had, no
doubt, been communicated to it where the
swell was greater. It was a singular thing to
see this great lump of ice thus navigating on a
voyage of its own, and not less interesting to
speculate on its origin and history. It had
probably been once much larger, and was now
consummating its own destruction, not only by
journeying to the warmer regions of the south,
but by its swinging motion, for the water some-

times washed right across its middle, where it
was low, and was thus actually sawing it in
two : its sides too were scratched and scored in
all directions, probably from knocking and
rubbing against other pieces of ice : it was in
some parts covered with snow, part of which
struck me as being rather discoloured and dirty,
but it had no signs of gravel or pebbles. The
discolouration might have been caused by the
presence of seals, as we had frequently seen
large patches of ice dirty and discoloured,
in spots where the seals had brought forth
their young, which are called by the men
" whelping-grounds." Sailed slowly during
the day, and in the evening sent out a punt or
two, which, just before dark, brought in an old
hooded seal, two old harp-seals, and several
young ones.

There was a beautiful aurora this evening,
exhibiting the usual arch or undulating band of
glancing rays, from the north-east to the north-
west, passing over both the Bears. The principal mass of light was in the north-east, and it
seemed to travel slowly towards the north-west.
To me the rays seemed to glance upwards perpendicularly from the earth, and the arched

appearance, as well as the radiation from a supposed centre, was probably due to perspective
and to the different angles under which the various parts met the eye of the spectator. The
light was fainter in the centre, and more full
towards each extremity of the arch ; and the
arch, instead of being perpendicular to the
earth's surface, like the rainbow, appeared
horizontal, or rather as if it were part of a
circle parallel to a plane which formed a
tangent to a point on the surface of the earth
some distance to the north of us. From such
a flat, circular, and sinuous band if rays be
supposed to glance perpendicularly upwards,
while an undulating motion travels along the
band itself, the form of the phenomenon will
be described as it appeared to me. I do not
feel myself competent even to speculate on
the cause of the phenomenon ; but it has since
struck me that if a tangent plane were to pass
through the magnetic pole,* and at some distance above this, but parallel to it, there were
a wavy or sinuous circular band of light, having a point above the magnetic pole for its

* Meaning, of course, the one situated just north of
America.

centre, and a radius of some ten or twenty
degrees of latitude or equatorial longitude,
with rays glancing upwards from the band
parallel to the magnetic axis of the earth, and
an undulating motion communicated to the
whole, it would produce the appearance most
commonly observed by me in the aurora seen
in Newfoundland. This appearance, however,
was by no means constant, a diffused yellow
light being sometimes alone observable.

March 23rd.—A furious north and north-
west wind with driving snow-storms. Sailed
under a close-reefed fore-topsail through some
open pools of water, and then, being brought
up in a compact field of ice, furled all sail and
moored to a large pan, where there were some
seals about us. The men picked up two or
three, but could not go far because of the blind-
ing snowdrift. The thermometer was down at
18° in the middle of the day. All hands snug
under hatches the whole of the afternoon.

March 24th.—Fine morning, clear, but cold,
with a light wind from the west-north-west.
There were six or eight vessels in sight. The
ice was rather loose and interspersed with lolly,
and the men were bringing in a few seals, but

had to go a long way for them. Among them
was the pelt of a female hood-seal and two
young hoods, one of which had not long been
whelped. As we were at dinner to-day we
heard an outcry on deck, and on running forward found a fine young shark lying on a pan
of ice alongside the vessel. A pile of seal-
pelts had been gradually collected on the pan,
which touched the bows of the vessel, and two
men were standing on it, helping to hoist
them on board, when the shark, attracted, no
doubt, by the pleasant savour, poked his nose
up over the side of the pan, and the men immediately hooked him with their gaffs, and
dragged him on to the pan without his making any great resistance. He was about ten
feet long, and he lay quietly enough, giving
merely an occasional lash with his tail, and
suffered us to roll him over and haul him
about with our gaffs. Stuwitz, of course, did
not lose the opportunity of sketching his form,
taking his dimensions, and cutting him up.
In his stomach we found the hinder part of a
young seal, part of a flatfish, and a small fish
like a gurnet. His liver was six feet long, and
occupied the greater part of his inside. Both

his eyes were dull and opaque, and very much
sunk in his head : they were apparently dis-
eased, and each had a parasitical animal adhering to it (a hernæopod), about an inch
and a half long, having two arms springing
from its body, that joined to form a small cup-
shaped sucker by which they were attached.
The structure of the shark's nose was very curious and delicate, having internally a series of
rows of filaments on a common base, something
like the gill of a fish. The five holes in his
neck to admit water to his gills, his elastic jaws
containing a fringe of teeth, and his extensive
gullet armed with small hooked teeth pointing
inwards, were among his best-known charac-
teristics. His flesh was beautifully white, and
the arrangement of its flakes and fibres very
curious and beautiful. We had some of it
dressed for tea, and found it not badly tasted,
but rather tough and dry, something like an
old halibut. He scarcely moved while being
cut up, but seemed almost as much alive after
having his inside cleaned out and being slit
open from head to tail, as he was before.
About an hour and a half afterwards, on cutting down through the head into the brain and

spinal marrow, he gave one or two vigorous lashes with his tail, after which there was no more motion. After taking what parts of it Stuwitz wished to preserve, we left his carcass on the ice, a prey to the seals, if they chose to accept of it. Of these latter we had brought on board to-day 356, making now 4186 as our total number on board.

March 25th.—Most beautiful morning, with a light breeze from the south-west. We lay nearly all day tightly jammed among strong ice, of which the whole surface, being covered with newly-fallen snow, was most dazzling under the bright sun. Several vessels about us, one of which, a schooner, was as full as she could hold, and was accordingly homeward-bound. A number of icebergs were in sight on all sides of us, none of them very near, and none of them in any way remarkable, except for mere peculiarities of shape or size. One, however, had a dark band, apparently of blue transparent ice, traversing its upper portion, in a nearly horizontal direction. I certainly do not understand in what way icebergs are formed, or how so a large body of compact solid ice can accumulate. Whenever this blue ice occurs in them, I take it to have resulted from a pool of fresh water having formed, either from rain or melted snow, on the surface of the iceberg, and being subsequently frozen, and then buried under other accumulations of white opaque ice. I question very much whether any mass of transparent ice is ever formed in sea-water ; and I am also inclined to believe that nearly all sea-ice is the result of snow saturated with cold salt-water and then frozen, rather than the actual congelation of the water of the sea. About noon the sky clouded óver, and we had a heavy snow-storm, with a gale from the north-east.

March 26th.—Fine, but cloudy ; thermometer at ten A.M. 32°. Made sail to the northeast slowly through thick ice. In the afternoon there was a silver thaw, that is, rain freezing as it falls, and covering everything with a coating of clear ice. Our masts, yards, and rigging seemed to be cased in glass. A young hood-seal being brought on board, Stuwitz drew, and measured him, but as it got dark, we put his carcass into a punt to dissect in the morning.

March 27th.—A gale from the south-west. On coming on deck this morning we could not find our seal, and on inquiry discovered that the middle-watch had eaten him during the night. The constant employment of the men on deck, when they had nothing else to do, was boiling, frying, or roasting pieces of seal flesh and eating them. Immediately after a dinner or breakfast down below, they would come on deck and set to work at the seal by way of dessert. Their constant food both at sea or on shore being fish or salt pork, fresh meat is at all times a luxury to them, even though it be that of a seal. It is amusing enough, however, that more than one half of the men going to the ice are Irishmen, and strict Roman Catholics, who would rather undergo any privation than eat meat on their fast-days, which in Newfoundland are Fridays and Saturdays. I had always found during the previous summer that my men, if there were no fish to be had, would confine themselves to bread and butter and tea on those days, even while undergoing the hardest labour. The good fathers of the church, however, either in pure ignorance of natural history, or by a little pious fraud, willing to indulge their flock during the cold and hardships of a seal-

ing voyage, have come to a unanimous determination that *seals are fish!* * I by no means
wish to disturb the consciences of any of the
men in so very harmless a matter, and hope
that seals may still be reckoned to be fish ; but
I am afraid I rather staggered one man by asking him if he ever heard of any fish that had
hot blood, and that suckled their young.   In
the mean time we determined to try what sort
of food a young seal was, and ordered one to
be cooked for dinner.   He was towed through
the water great part of the morning, then par-
boiled, and afterwards cut up and fried with
onions.   In this way it really was not bad : the
flesh was rather dark and strong, but by no
means so disagreeable as that of some sea-birds
I have eaten.

There was a fine aurora at night, consisting
of a bright undulating band just visible above
the horizon, with glancing rays shooting upward.

March 28th, 29th, 30th, and 31st.—Nothing
remarkable happened during these four days.

* I have, however, an idea that this determination is not
confined to Newfoundland, but that in the old rules of the
church, seals, otters, whales, porpoises, and all cetacea and
amphibia are classed as fish.

There was moderate weather the whole time, and one day we took off the skylight to let in some fresh air and let out the flies which annoyed us. We sailed about in various directions among loose ice, picking up a hooded seal or two now and then. Captain Furneaux says he never knew the field-ice so open and thin in this latitude (about 50°) as it was this year. There were very many icebergs about, some of them of large size, but in none could I see any signs of pebbles or boulders.

I saw one day four men attack a party consisting of a male and female hooded seal and their young one; they killed the female and the young one, and shot the male or dog-hood in the water, and hauled him on to a pan : when there, however, he recovered strength, twisted the gaffs out of the men's hands, and, in spite of what all four could do, he got away, and scuffled into the water, where he dived and disappeared.

The following is a summary of what I learned of the seals in Newfoundland :—

There are four species known on the coast. 1. The bay-seal. 2. The harp-seal. 3. The hooded-seal. 4. The square fipper.

1. The bay-seal*, as its name denotes, is confined to the bays and inlets, living on the coast all the year round, and frequenting the mouths of the rivers and harbours. It is the smallest of the four, and prettily marked with irregular spots of a small size. From what I heard, I am led to suspect that it breeds in the autumn or fall of the year. It is never found on the ice among the seals we had been pursuing.

2. The harp-seal† is so named from the old male animal having, in addition to a number of spots, a broad curved line of connected blotches proceeding from each shoulder and meeting on the back above the tail, forming a figure something like an ancient harp or lyre. The female has not this harp, neither has the male till after his second year. The young when born are covered with the white fur already described—they are then called " white coats:" at about five or six weeks old they shed this white coat, and a smooth spotted skin appears—they are then called young harps.

---

* This I believe to be the common *Phoca vitulina* of Linnæus, which is found on our own coast.

† This is the *Phoca Grœnlandica*, (Müller,) or Greenland seal.

When twelve months old the males are still scarcely to be distinguished from the females, and during that season they are called "bedlamers." The next season the male has assumed his harp. The harp-seals, as will have been gathered from what precedes, herd together, at least during the breeding season, and probably at other times. They are not seen on the coast of Newfoundland at other times, and probably come from the north to the ice-fields on the northern shores of the island for the purpose of bringing forth their young. The mothers leave their young on the ice, and fish about the neighbourhood for their own subsistence, returning occasionally to give suck. We did not absolutely see one suckling her young one, but we found the milk in the mouths and stomachs of one or two young ones that were brought on board, and it was of a thick creamy consistency and of a yellowish white colour. Meanwhile the males are congregated together in the open pools of water, sporting about. The young ones increase in size very rapidly from their birth, and are fattest at about three weeks old, at which time they are almost half the bulk of the old ones. From that time the fat diminishes slightly, although the bulk of the internal body increases.

3. The hooded-seals * are larger than the harps. Their skin is of a lighter grey colour, with many dark irregularly shaped spots and blotches of considerable size. The male, called a dog-hood, is distinguished from the female by the singular hood or bag of soft flesh on his nose. When attacked or alarmed, they inflate this hood so as to cover the face and eyes, and it resists seal-shot. It is impossible to kill a dog-hood, even with a sealing-gun, when he has either his head or his tail turned towards you; and the only way is by shooting him on the side of the head, and a little behind it, so as to strike him in the neck and the base of the skull. The young of this species are not provided with the thick woolly coat of the young harp-seals, or if they are it is shed very shortly after birth. They have whitish bellies and dark grey backs, which when wet have a bluish tinge, whence they are called " blue-backs." Those which were brought on board alive seemed much tamer and more gentle than the

* This species is the *Stemmatopus cristatus* of F. Cuvier. (See the article SEAL in the ' Penny Cyclopædia.')

white coats, and when teased did not offer to
scratch and bite so much as the others.   Their
fat is not so thick, and they are consequently
of inferior value.   The hooded-seals do not form
such large herds as the harps, and the male
and female seem to keep more together, both
being commonly seen near the young one.
The hooded-seals generally bring forth their
young two or three weeks later than the
harps, and they always occupy different districts, the hoods being generally found farthest
to the north.   In the stomach of the old
hooded-seal which we opened we found nothing but portions of squids, and two or three
undigested beaks of large cuttle-fish.

4. The "square fipper" * is described as
being much larger than the hooded-seal.   It
is, however, very rare; and we did not see one
or hear of one being seen this season.

The morse, or sea-horse, is occasionally met
with and killed by the seal-hunters, and not
unfrequently a white bear or two.   We only

---

* I do not know what is the scientific name of this seal,
or whether he has been described at all.   Captain Furneaux
told us they were sometimes twelve, and, if I recollect
rightly, even fifteen feet long.   It may possibly be the *Phoca
barbata* of Müller.

heard of one of the latter being seen this year, and it was fought some time by the men with their gaffs, dodging it from pan to pan, till at last one or two, coming up with their guns, shot him. We did not hear of this, however, till after our return to St. John's. White and other foxes are also occasionally found on the ice.

There is no certain rule to be adhered to in searching for the herds of seals, the great secret of success being constant activity and pushing about through the ice till they are met with, making for the strongest and largest fields of ice, and watching the set of the currents, the winds, and other circumstances which are likely to lead to them.

April 1st.—Thick and foggy, wind south-west; sailing to the south-east through bays of water. Saw a punt's crew of a schooner near us kill a family of hooded-seals, consisting of the male, female, and young one. Distinctly observed the male blow up his hood as the men approached. The hooded-seals are more fearless, and show more fight when attacked, than the harps: they will, however, always get away if they can. The men say that if they can once kill the female, they are sure of the rest, as the young one does not stir, and the dog will not go far from the spot, but keeps continually popping his head up in the holes and pools about, growling and whining after his mate.

It was quite a pleasure to have a little clear sailing this morning, after being so long stuck fast in the ice. The thermometer at nine, A.M., showed 35°, and in the middle of the day the fog cleared off, and we had beautiful weather. We moored the vessel to a large pan of ice, and filled all our water-casks from a pool of clear water in its hollows, taking the opportunity also of getting a good wash ourselves, a luxury we could not very often enjoy. There were several vessels about us, two of which had not seen a single white-coat this season, so capriciously scattered are the herds of seals. There was a very singular island of ice a few miles to the south of us. It rose to a height of about one hundred feet, and resembled the worn and ruined cone of a volcano. At about half its height was a platform sloping to the north, and round three of its sides a thin wall of ice rose up sloping inwards, and was pierced

towards the south by a great arch somewhat
resembling a Gothic window. The sloping platform was strewed with blocks of ice, fragments,
probably, of its surrounding walls. As it was
in the direction in which we intended or expected to proceed, we hoped to get a nearer
view of it. The crew amused themselves this
afternoon with jumping-matches on the snow,
and dancing reels on the ice, to which I added
darting at a mark with the gaffs, till the captain put a stop to that part of it by begging
us not to spoil any more of his gaffs. By an
amplitude observed at sunset this evening with
my prismatic compass, I found the variation
to be as nearly 38° as possible. The sunset
was clear and beautiful, lighting up the icebergs with hues of liquid gold and rose colour;
and at night the aurora was very fine, having,
in addition to the appearance I have already
described, a diffused yellow light over the
whole heavens, except the south. On looking
at the zenith there appeared to be a slight radiation from, or rather an apparent convergence
to it, produced by perspective.

April 2nd.—A most lovely morning, with a
light breeze from the west. Unmoored and

made sail slowly through the ice to the south, with the temperature at 33°, at 10 A.M. The arched iceberg is now south-west of us, and we can see its other side; the arch now appeared to be round below as well as above, and below it there floated a larger piece with a projecting pinnacle, that may probably have been detached from some part of the iceberg. It was still three miles from us, and the ice intersected with many small channels and lakes of water, rendering an attempt to reach it on foot highly dangerous, as the ice frequently separates and breaks up without warning given, and even a narrow channel might cut us off from the vessel. Taking a punt would be hazardous, for the opposite reason, as the ice might close in and oblige us to abandon it. The latter predicament was one in which a punt's crew and myself were nearly caught this afternoon. We had gone out where the ice was much divided into channels and pools of water, but it gradually closed in upon us, and we had much difficulty in forcing our way back to the vessel, being obliged to get out on opposite sides of the boat and force the pans asunder with our gaffs to make a passage for her. We brought a young harp on board alive, with its new coat on. Tying a line round one of its hinder fippers, we let it swim about near the vessel, which it did with great grace and agility, diving and rising again to the surface, and occasionally crawling out on a pan of ice. The beautiful clearness of the water enabled us to watch all its motions. It swam like a fish, principally by the motion of its body, closing and spreading out its hinder fippers into the exact form of the tail of a fish. Nestor jumped over into the water and swam up to it; but when the seal saw him approach it raised itself upright in the water, and when within reach attacked him tooth and nail, biting him severely in the lip, and making master Nestor turn tail immediately. Their heads and bodies are so round and smooth, that a dog cannot seize them anywhere except in the fippers.

April 3rd and 4th.—Continued our route to the southward slowly, the ice becoming more open and broken. Many icebergs were still in sight, and several vessels bound homeward like ourselves. In the afternoon of the

4th a thrush (turdus migratorius), very common in Newfoundland in the summer, and called there a blackbird, though it has a brown body and a red breast like a robin, flew several times about the vessel, hopping about and appearing lively and in good condition. The men said they had sometimes seen them in flocks on the ice in the beginning of April, and that two or three would often come on board and stay with them till they neared the land, picking up the crumbs and making themselves quite at home. This evening the lakes of water became wider, and as we sailed to the southward the swell gradually increased, till during the night I was awakened by the vessel rocking as if in the open sea.

April 5th.—A very heavy gale blowing from south-west. We hove the vessel to, under the lee of a thin skirt of ice, and I witnessed the sea in the grandest aspect I think I ever saw it. There was no ice to windward of us but this thin skirt about a mile broad, which was marked into small round pieces about a foot wide, forming a perfect mosaic pavement on a gigantic scale. This was sufficient to prevent the formation of the numberless flickering waves and crests of foam and spray into which the surface of the sea is lashed by a gale of wind, but it offered no impediment to the swell. The unbroken swell of the Atlantic ocean accordingly rolled in upon us in huge continuous ridges, heaving the pavement of ice on its mighty folds, and alternately lifting us on its broad domes and swallowing us in its deep hollows. The absence of all minor waves made the primary undulations to be felt in their true magnitude, and it was certainly a magnificent sight. Grand as was the sight, however, I would in a short time have gladly exchanged it for a station on the solid ice again, as both Stuwitz and I began to be sea-sick.   More than half the crew, I found, followed our example, and for the next day or two barely enough of the watch were left on deck to manage the vessel.

April 6th.—When the wind moderated a little, we made sail to the south, and in a short time left all the ice behind us.   At noon we were in lat. 50° 41, supposed longitude 50°, with the wind at west-south-west, apparently moderating.

April 7th to 11th.—One continued gale from south-west to north-west, with very heavy sea running; top-gallant yards sent down on deck, and with close-reefed fore and mainsails, &c. trying to make what head we could by beating up to the westward, but the wind constantly turning against us on both tacks. By soundings, we found ourselves to be on the banks, and supposed we were on the inner or western edge of them. We accordingly ran down as far as lat. 47°, and then tacked at 47° and 49° alternately, St. John's being in 47° 33', and the current supposed to set to the southward. Meanwhile, our seals began to run, the pumps brought up morning and evening a lot of thick white stuff like syllabub or soap-suds, the mixture of the seal-oil and salt bilge-water, being agitated in the pump. The ship accordingly was perceived to stink most awfully, and everything on board, including the bulk-heads of the cabin, began to sweat with grease. Our clothes, also, became smooth and polished, especially where there was any pressure. Our fresh stores had long been exhausted, as also our wine and spirits, and we were now reduced to tea and the common rum, and had not much of that left. At last, in the afternoon of the 11th, the wind moderated and shifted into the north, the sea

went down, and we got a fair run to the westward, and at sunset some of the men thought
they could see Cape St. Francis in a bank of
fog to the northward, when it fell nearly calm.

April 12th.—About three o'clock this morning we were awakened by the happy intelligence
that the lighthouse of Cape Spear was in sight,
and coming on deck at daylight, we had the
pleasure of seeing the bold and barren coast
of Cape Spear and its neighbourhood about
fifteen miles from us.   The snow had all disappeared, except in the hollows and crevices of
the cliffs.   About eight o'clock a light breeze
sprung up from the south-west, that just sufficed to carry us into the narrows before noon.
We found the harbour full of vessels, and the
first boat that boarded us brought us the intelligence that many vessels had come in with
five, six, and even seven thousand seals on
board ; that some had sailed a fortnight ago on
their second trip ; and that seals were in consequence down as low as 15s. the cwt.   This
lengthened the visages of our crew perceptibly,
they having expected 18s. a cwt.   Stuwitz
and I were, however, too busy in landing and
getting away to our lodgings to trouble our-

selves about matters that did not concern us; and when there, we bathed first in hot water and then in eau de Cologne, ordering all our clothes to be hung out of doors, as we observed that the acquaintances whom we met as we passed through the streets studiously got on the windward side of us; and I have no doubt had we been turned out to be hunted, we should have left a " burning scent " behind us.   Upon the whole, we were highly pleased with our expedition to the ice, which we could not have seen under more favourable circumstances.   It was a very good season : one vessel in two trips brought in eleven thousand seals ; and the total take this year must have been considerably upwards of five hundred thousand.

END OF VOLUME I.

London ; Printed by WILLIAM CLOWES and SONS, Stamford Street.